ENERGY MATTER AND FORM

TOWARD A SCIENCE OF CONSCIOUSNESS

PHIL ALLEN, ALASTAIR BEARNE, ROGER SMITH

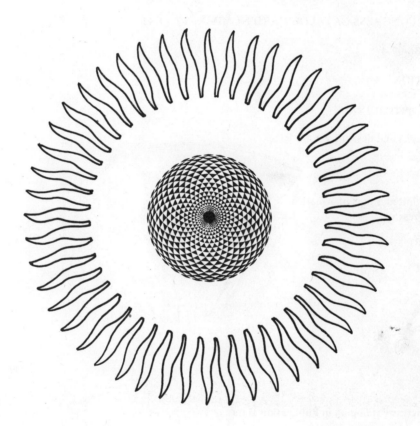

Volume II in the six volume series:

"THE SUPERSENSITIVE LIFE OF MAN"

CHRISTOPHER HILLS, GENERAL EDITOR

UNIVERSITY OF THE TREES PRESS
P.O. Box 644
Boulder Creek, California 95006

Copyright © 1977 by University of the Trees Press

All Rights Reserved. No part of this book may be reproduced in any form or by any electronic or mechanical means including information storage and retrieval systems without permission in writing from the publisher, except by a reviewer who may quote brief passages in review.

LIBRARY OF CONGRESS CATALOG CARD NUMBER: 77 - 84873

ISBN: 0-916438-07-4

SECOND EDITION

First Edition Copyright 1975

Cover photographs and design by John Hills

Printed in the United States of America
by Kingsport Press

Library of Congress Cataloging in Publication Data

Allen, Phil, 1944-
   Energy, matter, and form.

   (The Supersensitive life of man; v. 2)
   Bibliography: p.
   Includes index.
   1. Consciousness. 2. Science–Philosophy.
I. Bearne, Alastair, 1948- joint author. II. Smith, Roger, 1953- joint author. III. Title. IV. Series.
BF311.A48 1977   128'.2   77-84873
ISBN 0-916438-07-4

To that inner spiritual sun, source of us all --
The One who sees through every eye.

ACKNOWLEDGEMENTS

Participating in the birthing of this book was a real growth experience for both of us. We express deep appreciation and heart-full thanks to our family at the University of the Trees, who have helped us grow and have given so much of themselves in making this book a living reality. When we needed help, help appeared. Through long hours of proofing, editing, typesetting, illustrating, etc. this group of selfless servers was always there with helpful criticism, words of encouragement, smiles and hugs. At times it felt like we had forty arms and forty legs! This book, then, is much more than the creation of the authors. It is a manifestation of group consciousness in loving action.

We give special thanks to: Pam Osborn and Debbie Rozman for many long, patient, and devoted hours of typesetting; Ary King for her invaluable help in editing and indexing; Michael Hammer for his fine photographic work; Gary Buyle for his imaginative illustrating of the section and book title pages; John Hills for the strikingly beautiful cover; Allan Allen for his creative illustrations which appear throughout the text; and Christopher Hills for the inspiration to write it in the first place.

We also express our appreciation to Alastair Bearne whose first edition of the book provided a firm foundation for this one, to Earle Lane for the electrophotographs in Chapter One and to all those we have drawn upon for information and ideas.

To all of you, thanks.

ABOUT THIS BOOK

This book is a cross-disciplinary overview of many years of research into the nature of man by dedicated people in many branches of science. Attention has been directed to the latest developments and discoveries in several new fields, the psychosciences, with concentration on those most directly concerned with the development, or evolution, of human consciousness. Most notably, we have tried to introduce the theory of conscious evolution developed by Christopher Hills, D.Sc., in NUCLEAR EVOLUTION and SUPERSENSONICS. With the recognition that our awareness is just as real as the physical things we experience with it, has come the realization in many of us that consciousness is a basic ingredient of the universe and scientific theory must now be revised to include it.* In this new paradigm Christopher Hills offers us a valuable map for exploring the structure of our consciousness as well as developing the new tools needed for investigating those subtle dimensions of energy which cannot be studied by current scientific methods.

Aware people in nearly every branch of science are beginning to include consciousness in their theory and experimental research. As a result, radical transformations are underway in many sciences due to a major *paradigm change*** (called changing world-view in this book) toward a new macrotheory which brings consciousness into cosmology, biology, physics, etc. Historically, in seeking truth in materiality, science as a discipline was split into many branches with no unifying principle (except for the mistaken assumption that reality is a purely physical phenomenon, hence the dominance by physics over the other sciences). But now the sciences appear to be converging and merging into one unified science of consciousness as more of us realize its fundamental importance. Consciousness just might constitute that universal field which Einstein was trying to establish at the time of his death. (After all, we all have it.)

There is a growing movement underway in physics to see the universe as a mental phenomenon as well as a physical one. In subatomic physics, for instance, the concept of solid matter is changing. Instead of matter and empty space the universe now appears more like a dynamic web of inseparable energy patterns† which can be influenced by our thoughts.††

---

* Note the observations of cosmologist Arthur Young, page 128, and Nobel Prize Physicist Eugene Wigner, page 90.

** See Kuhn, "The Structure of Scientific Revolutions."

† Note observation by Dr. Fritjof Capra, page 6.

†† Note research of Dr. Harold Putoff, page 105, Professor I.M. Kogan, page 102, Dr. John Jungerman, page 111, and Dr. Victor Inyushin, page 112.

These energy patterns manifest as wave vibrations* which form interference patterns according to the principles of holography. Physicist Jack Sarfatti reasons that the entire universe may function like a giant hologram (page 121). This holographic theory of reality is giving birth to the new science of psychophysics (our term) called the "new physics" by Heisenberg, Sarfatti, Wheeler, Wolf, et al. This new science includes the consciousness of the participating scientist in the theory. And this is a key point from the standpoint of the psychology of science.

A basic tenet of the scientific method is the principle of the objective observation where the "subjective" element is supposedly eliminated through careful control of the variables and the requirement for replication by any competent observer. However, the biggest variable of all is the consciousness of the observer. As we point out, scientific proof is supposed to be *self-evident.* Yet, because the "subjective" has been ruled out in the objective search for truth, how can there be any self to refer that evidence to? Experiments with interferometers have shown that they are sensitive to the mere presence of someone in the same room. According to quantum mechanics a detached observer is impossible. Physicist John Wheeler feels that it is time to replace the observer with the participator, one who openly and honestly takes into account the heretofore hidden variable, the quality of his/her own consciousness. This new principle in the scientific method unites the subjective with the objective, the psychological with the physical, hence the term psychophysics.

From here the next logical step is to investigate human perception. So we consider the work of Professor Vasco Ronchi and the "new optics" (page 95). We find a new application of relativity in that perception is found to be relative to the consciousness of the perceiver (as psychological tests with drugs such as LSD have clearly shown). In his research data, Ronchi thoroughly demolishes classical explanations for how we see** and provides simple experiments by which everyone can see that perception is essentially psychological, not physical. *This is the key to the paradigm shift, the major thesis of this book.* The work by Ronchi is crucial for it allows us to prove to ourselves that classical explanations for how we see (which we have all been taught) are based upon hidden assumptions, rigged experiments, and mathematical explanations which

---

* Which we have charted as the electromagnetic spectrum. However, this spectrum probably extends to zero on the lower end of decreasing vibration and extends into infinity at the other end.

** Not proven conclusively in our books perhaps, but in his original work, much of which is covered in great detail in "Supersensonics" (a companion volume in this series).

have little or no correspondence with reality.* A few of Ronchi's less complicated experiments are included so that anyone can prove that what is seen outside is actually seen inside -- inside our consciousness (if we are not conscious we cannot see, so perception is dependent on consciousness). This turns the paradigm of science inside-out. No longer is there a material world "out there" separate from the observer. It is all a mental world "in here," inside the consciousness of the participator.

This leads to the idea that there may be other ways of sensing just as valid as our normal five senses; many scientists have made their greatest discoveries thanks to what they call a "sixth sense," or intuition, or creative imagination.

In Section One, the same convergence of the physical and mental is seen in psychophysiology. Chapter Two deals with the up until now esoteric subject of energy centers in the human body, their resonance with certain octaves of vibrations corresponding with the seven colors of the rainbow spectrum, and a typology of personality based on levels of consciousness associated with differing inner psychological time worlds. These psychological attributes are found to have real physical effects. As a result, psychology is joining physiology to bring the new science of psychophysiology.

A similar synthesis is occurring in other sciences bringing cross disciplines such as psychobiology, psychochemistry, psychoneurology, etc. The scientists in these fields are engaging in lively cross-fertilization of ideas which are bringing startling revelations about the true nature of reality.

---

* By staying observers and not checking out the classical explanations we have believed what we have been taught. As participators, however, we can check things out experientially and know what the truth of the matter is directly.

INTRODUCTION
TO THE SECOND EDITION

Since writing the introduction to the first edition two other students of consciousness at the University of the Trees have taken Alastair Bearne's accounts of the work done prior to 1970 at our London center and brought them completely up to date. Phil Allen and Roger Smith undertook this responsibility and acknowledgement is due to their many hours of loving labor in re-writing the book for a 1978 production to include current interest in black holes and pyramids, etc. Credit is also due to the typesetting ability of Pamela Osborn and Deborah Rozman who dropped everything else to get the book into print for those many people who patiently waited for the new printing.

Since the first printing of this book as Volume II of the series "The Supersensitive Life of Man" the University Press has printed the first edition of Volume III called *Supersensonics* which has now almost sold out. *Energy Matter and Form* is the perfect introduction to this new science which has found such a following amongst those at the forefront of consciousness research. In this much expanded edition the reader will find almost a completely new book. Therefore I have to thank the three authors concerned for their dedication to the "Supersensitive Life of Man." No doubt this book will be brought up to date frequently as it is used as a text book in the teaching of parapsychology. Anyone who has a new discovery or keeps any careful records of results either on pyramid research or other psychotronic aspects of *Energy Matter and Form* is welcome to submit it to the editors for inclusion in future editions.

Christopher Hills
Boulder Creek, California

INTRODUCTION

TO THE FIRST EDITION

It was in 1970 that I asked Alastair Bearne to review the various aspects of the research covered by our laboratory at Centre House, London. Using a wide variety of resources and the tapes from my lectures and papers collected by The Commission for Research Into the Creative Faculties of Man, he has given us an outline of some of the areas covered by a few of our members throughout the world.

When this Commission was founded in 1962 it was my intention to gather together a cooperating group from that little known band of students who quietly reach far beyond their contemporaries into the hidden forces of nature. Vaguely they know of each other's existence and sometimes they meet and share their advanced knowledge of human consciousness. However, I found in actual practice that although yogis and occult students may be very advanced as individuals, and may have acquired personal powers far beyond the ordinary comprehension of man, there was still very little attempt at evolutionary and unifying techniques in terms of group advancement.

These highest masters in their own fields, who knew more than they could ever use or ever communicate to fellow men, were unable or unwilling to put aside their own individual enlightenment and come together as a group. This resulted in my having to supply most of the research materials for the Commission when the real intention was to unite our forces in love and wisdom. Hence it was necessary to commence the Centre project and experiment in the evolution of the group consciousness at Centre Community in London. Developing the leaderless leadership and the study of methods of self-government and self-confrontation seemed to me to be the first requirement even above the acquisition of special knowledge of the unseen worlds of spirit and nature. Most of the students who came to Centre House in the first years were taking a psychology degree and when Alastair Bearne had obtained his degree I obtained a postgraduate grant for him to review the research, under my direction, at the Centre. It is unfortunate that he could not stay longer to complete the work on further volumes so urgently needed. This need is now being filled by several other students, each of whom is tackling a separate aspect of a vast field of study – little of it publicly known and mostly confined to a small number probing human sensitivity at the fringe of human consciousness.

Our next and third volume in this series of "The Supersensitive Life of Man," will deal in greater detail with some of the aspects briefly mentioned by Alastair Bearne in this second volume. Therefore it should be considered, like the first volume, as more of an introduction to the far reaching knowledge of human potentials which are now being documented for the public domain.

The subject of radiesthesia, which this volume introduces, has been the study of scientifically trained people for some 50 years since the 1920's but it is still not a universally accepted method of investigating facts. However, used in conjunction with ordinary scientific methods, it can open a direct window into nature's invisible worlds. In a talk which was intended to last an hour at the British Medical Society for the study of Radiesthesia in London, I discussed the inhibiting of the effects caused by the misunderstanding of the scientific method. The group of doctors who were there pressed me to go on and the chairman invited me home to discuss the further implications. This meeting could not happen in American medicine where there is a fanatical opposition to anything which smacks of invisible or subtle energy which cannot be sensed by all.

There is a problem in that the field is open to so many quacks and seems to attract the half-baked scientist who lacks the rigor and skill needed to validate evidence. On the other hand there have been a number of brilliant scientists who have used this method of discovering new pathways without telling their more hidebound colleagues how they got their leads.

The old water divining faculty, clothed in new garments and used to examine all influences ranging from minerals and atomic substances to disease patterns and ion transport, had to keep submerged underground until scientific man discovered through the new technique of biofeedback that the nerve pathways of one's hand and neuro-muscular reflexes can easily amplify the smallest signals of perception when combined by a particular sensation in the mind. This sensation or feeling was once limited to the trained operator of divining instruments, but nowadays anyone can reinforce these controls over involuntary processes with sophisticated biofeedback machines. For all radiesthesists it would be logical for the scientifically trained investigator to become proficient and to enrich the biofeedback field with man's simplest and yet most accurate biofeedback machine – his own nervous system. This bioenergetic sensation which the radiesthesist waits for as a signal from the source is no different from that experienced through biofeedback conditioning or auto-feedback. This has always been practiced by yogis without using the more sophisticated instruments of today. The difference was only one of understanding the phenomenon and the validation of a meaning given to the perceptions. The operations of radiesthesia are as old as the human race and the writing of such books as the *I Ching,* the construction of ancient buildings such as Stonehenge and the Pyramid of Giza and the digging of wells and the forecasting of weather, are only a few of man's attempts to use man's greatest and most creative faculties in investigating nature's ways.

Radiesthesia and its various methods of sensing are no more miraculous than the homing instincts of pigeons and dogs or the mating selections of butterflies and moths, or the direction-finding of the salmon returning to lay their eggs. The study of divining phenomena shows why storks never nest over an underground stream and why anthills are almost always over one. The ancients who had no scientific equipment as we know it today knew all these things by other means and therefore were able to investigate fields of enquiry which are still considered unscientific by some today. All this is about to change as chemistry and physics realize that the exact sciences do not give wholistic pictures through measurement in arbitrary human units. Radiesthesia uses nature's own yardsticks and allows the various emanations and signals from an object to describe that object in its own vibratory language.

It was in 1920 that the first important congress of engineers, physicists and natural scientists was held. There were about 30 pioneers who had been looking into this phenomenon since 1913, among them Henri Mager and Dr. Vire, who later became director of the Paris Natural History Museum. Later M. Larvaron, Professor of Agricultural Chemistry at Rennes University, with Dr. Regnault of Toulon and Dr. LePoiree of Nice investigating the medical side of radiesthesia, exchanged findings with a civil engineer by the name of Louis Turenne who was a lecturer in radio propagation at the Fontainebleu Military Academy.

From here on hundreds of Frenchmen joined the movement resulting in about 500 books mostly in French. Dr. E.A. Maury, a student of Turenne who became one of the first medical practitioners to use radiesthesia, invented along with Turenne a great number of methods of using rods and pendulums.

The British Society of Dowsers was a group of eminent people who formed an association in 1935, and in 1938 a group of medical doctors, who had been to France to study, helped to establish The Medical Society for the Study of Radiesthesia in London. This is the group of medical scientists that I had the privilege of addressing at the headquarters of the British Medical Society in 1966, on the difficulties that radiesthesia presented to the scientific method.

Mrs. Dudley Wright F.R.C.S. and Dr. Guyon Richards, who had both helped to found the M.S.F.S.R. in 1938, encouraged Noel MacBeth to start a teaching centre which would translate the French works and make them available by correspondence throughout the British Isles. Two professors from the French Academy of Sciences, Dr. Branley and Dr. d'Arsonval, supported the work along with a considerable number of scientists who could not reveal their interest openly because of bitter condemnations by colleagues.

However, the president of the Medical Society for Study of Radiesthesia, Dr. McCready, whom I met at that lecture, befriended our Centre in London and gave several lectures on the subject of Homeopathy and its relationship to radiesthesia.

I cannot remember exactly how or when I met Noel MacBeth. For some years I had been using methods of divination to study occult phenomena and to test ancient wisdoms, in order to separate accurate knowledge from the mumbo-jumbo that often goes under the label of secret knowledge. From 1957 I had been receiving and had worked with the translations of French books by Noel MacBeth, in my mountain-top laboratory in Jamaica while investigating the nature of light and color. In 1960 we met at his house in the country and decided that I would distribute the Radiational Physics Notes, as they were then called, in America and the Western countries.

When Noel died he bequeathed all his stencils, notebooks, and translations of about 150 books to me and over the intervening years I have had various students working on them, with the aim of eventually bringing out a clear, concise and logically ordered volume of these techniques. We also aim to establish a museum with all the old rules and discs which were used in the early days to detect influences.

The fact that these influences are vibratory, and that a pair of substances of the same vibration are somehow linked throughout space even when solid matter is intervening, was revealed by experiment to depend on some intermediary force of nature which was originally defined as a carrier wave. The sun's rays and the earth's magnetic field were found to be carriers as well as rays from Turenne's "Radium Block." This led investigators to postulate some subtle etheric carrier which Einstein referred to as the *universal field* and Turenne called "phenomenon D." Whatever the labels it did not matter, the results were there for anyone who was willing to become proficient in the recognition of diviners' signals.

Since all this material began to come into my hands over the last 20 years, I have seen many scientific verifications of the brilliant early investigators. In recent books on psychic phenomena and pyramid energy behind the Iron Curtain and the properties of mummification, dehydration, crystallization, etc., one sees clearly the work of these early invistigations, such as Bovis and Turenne, all of which had been completely published and documented years before. These results were confirmed by thousands of radiesthesia students and water diviners from 1920 onwards and are now being written up as new research and new knowledge by popular writers without any credit to the originators. Therefore the much more specific and detailed notes and books entrusted to me are being rewritten and published with the translations of the original authors' results by students professionally trained in physics and engineering. They are working to simplify the connections with standard physics and the relationship to radio propagation and will bring out a later and separate volume for the use of those who wish to develop their potentials in the deep study of what we have re-named "Radiational Paraphysics."

This is an exciting study for those who can accept that the human race has already the equipment in the frontal lobes of the brain for the acquisition of knowledge which was formerly only available to seers, sages and divine incarnations. So much does this method of Radiational Paraphysics reveal, that psychic feats, such as the miraculous draught of fishes by Christ, can become commonplace, brought within the powers of anyone who will train his or her hand to sense a shoal of fish under the water. Just as the water diviner's twig will detect water under the ground, so is it possible to detect fish or gold or the whereabouts of any lost objects.

On one occasion in France, when visiting Picasso, I had my passport stolen and lost traveller's checks and irreplaceable papers and later recovered them while staying as a guest of Eileen Garrett, the founder of the Parapsychology Foundation in St. Paul de Vence several miles away, all by using these methods. I was able to tell the police exactly which house they were in and which person stole them and all this is documented. Therefore we are dealing here with ways of knowing which are supra-mental and capable of revealing more about nature than any university education or any library of information or any computer devised by man.

You would think that scientific man, the intellectual phenomenon of intelligence so valued in our times, would jump at such a chance of having a direct perception of nature's workings so as to compare and validate and make it consistent with what is called "science." But over the last 25 years we have found this not to be true. By some strange cosmic quirk the secret knowledge of the universe remains limited to those who have the insight to see what they are getting. This knowledge, I have found to my own astonishment, cannot even be given away or shouted from the housetops. The reason is obvious but still a cosmic riddle. A man cannot know what he does not know. If a man knows what he does not know he is certainly a wise man. These volumes in this series can only be shared with those who have the insight and the openness to try to become proficient in radiesthesia before making any judgements about it.

CONTENTS

Introduction by Christopher Hills   *i*

       SECTION ONE: ENERGY AND CONSCIOUSNESS   *by Phil Allen*

INTRODUCTION   2

CHAPTER ONE   **WHO ARE WE?**   6

Energy Systems and the Human Body – Energy Fields and Life – Our "Bioenergy Body" – A Phantom Revealed – A Cancer Detector? – Acupuncture Research – We Are Light

CHAPTER TWO   **OUR RAINBOW BODY OF LIGHT**   19

The Etheric Aura – The Center Concept – The Chakra Energy Centers – The Physiology of the Chakras – Health and Eye Light – Seven Colors and Seven Levels of Consciousness – Red Physical Level – Orange Social Level – Yellow Intellectual Level – Green Security Level – Blue Conceptual Level – Indigo Intuitive Level – Violet Imaginative Level – What Lies Beyond? – Experiment Section – Sensing the Etheric Aura – An Aura Detector – Sensing the Chakras – The Aura Pendulum – The Spectrum Mirror Pendulum – Concentration

CHAPTER THREE   **HERE COMES THE SUN**   52

Life Energy – Prana – Qi or Chi – The Fifth Element: Life The Sun is on Fire – Our Body is on Fire – We Breathe Fire – Our Blood is on Fire – Our Cells are on Fire – How We Breathe Fire – We Are Giant Electromagnets – A Symbol of Life's Fire – We Are Dynamos – The Fire This Time – Kundalini: The Sun Within – The Kundalini Roomph Coil – Seven Stages of Growth – Nuclear Evolution – Experiment Section – Balancing the Breath of Life – The Abdominal Lift – Breathing Exercises – Rumf Roomph Yoga – Centering – Shutting Off the Senses – References for Section One

             SECTION TWO: MIND AND MATTER   *by Phil Allen*

INTRODUCTION   84

CHAPTER FOUR   **SEEING THROUGH MATTER**   87

The Changing World View in Physics – Relativity – Subatomic Physics – The Ideal I – Radiational Paraphysics: Supersensorics – The Open Mind – Cyclops: The Ideal Eye – How We See: Psychoptics – Our Mind's Eye – Our Mental Blinders – Experiment Section – Mirror Gazing

CHAPTER FIVE   SEEING THROUGH MIND   102

Psi Communication – Parapsychology – Extrasensory Perception – Psychokinesis – Breathing and Telepathy – Telepathy and the Weather – Psi Plasma – Psychotronics – Holography – The Holographic Brain Model – The Holographic Body Model – The Biologogram – Gravity and Space/Time – The Black Hole – The Superuniverse – Superspace – Consciousness and Reality – Direct Perception – At the Zero Point – Experiment Section – A Psychotronic Energy Detector – Telepathy Experiences – The Merkhet and Osiris Pendulums

CHAPTER SIX   SUPERSEEING   137

The World Culture – Planetary Being – The Evolution of Consciousness – The Psychosphere – Our Planetary Nervous System – Communicating With Light – The Plasma Crystal Screen – The Computer World Brain – The Holographic Imagination – A New Language – Psychosociology – Nature's Model of the Nucleus: The Living Center – The Positive Center Principle of Creative Conflict – World Government – Manna from Heaven – The Peaceful Center – A Constitution for World Peace – The World Peace Center – References for Section Two

SECTION THREE: VIBRATION AND FORM   *by Roger Smith*

INTRODUCTION   166

CHAPTER SEVEN   THE VOICE OF CREATION   170

Vibration and Form – Vibrating Forms – Our Vibrating Consciousness – Sounds of Rebirth – The Breath of Union – Healing With Sound – Seeing Sound – Making a Chladni Plate – Create Forms With Your Stereo – Crystal Oscillations – Defying Gravity – Liquid Forms – Moving Sculptures – Seeing Speed and Wavelength – Energy, Vibration and Form

CHAPTER EIGHT   RAYS OF LIFE   192

The Great Pyramid – Pyramid Energy – The Ray of Life and Death – The Amazing Coffer – A New Order for a New Age – Experiment Section – Excessive Claims – Building a Pyramid – Mummifying Food – Sharpening Razor Blades – Test Your Plants – A Cure for Negative Green

CHAPTER NINE   CREATIVE IMAGING   207

Real Images – Wholistic Health – Magic – How Do We Know? – Time is Not Real – The Inner Seer – Learning a New Language – A Question of Blindness – Distant Seeing – A Clear Mind – The Power of Inner Sleep – Dorje: A Symbolic Gateway to Health – Using the Gate – Awakening Our Inner Vision – The Magic Mirror of Life

CHAPTER TEN   SUPERSENSING   231

Radiesthesia: The Supersense – Concentration and Receptivity – Our Fantastic Antenna – The Yes and No of It – Aura Balancing – Construct a Simple Pendulum – Tuning Your Pendulum – Learning Its Language – The Dowser's Rod – Mapping Energy Fields – Building a Vocabulary – Electricity and Magnetism – Symbols are Maps – Colors of Life – Beyond Our Senses

CHAPTER ELEVEN   THE HEALER WITHIN   248

Our Vibrating Signatures – A Diagnostic Breakthrough – Healing Vibrations – The First Broadcaster – The Stick Pad – Color Healing – Tuning In – Radionics Works Without Electrical Circuits – The Real Circuit is in Our Consciousness – Consciousness Makes the Circuits Work – Belief Systems – Believing is Healing – Belief Systems and the Levels of Consciousness – The Healer Within – Radionic Broadcasting – Fail-Safe Broadcasters – The Square Balance Equalizer – The Magnetron – The Ultimate Instrument – References for Section Three

EPILOGUE   TOWARD A SCIENCE OF CONSCIOUSNESS   284

Futures Research – The Sciences are Merging – The Psychosciences – Consciousness Research

Index   290

SECTION ONE

WHAT IS CONSCIOUSNESS ?

This question is swelling like a tidal wave in the turbulent ocean of feeling and thought flowing through us all. Consciousness is something we all have, yet we feel and know it not. We seem to pay more attention to the *contents* of consciousness than to the source or origin of consciousness – *consciousness itself*. Consciousness goes beyond the limits of our imagination, for it is the essence of imagination itself. Thoughts are not consciousness, because we can be aware of our thoughts. We can even experience awareness without thought. If consciousness is the means whereby we know what we know and do what we do, then it is the fundamental ground of all existence and cannot be described in terms of anything else.*

Albert Einstein's famous and elegantly simple formula, $E=MC^2$, reduces everything in the universe to a single concept -- energy, or light. As we learned in Volume I energy is essentially vibration. Atoms vibrate, cells vibrate, *we* vibrate, each with a "note" in the "Cosmic Chorus" unique to ourself. The realm of physical vibrations of light is mapped by the electromagnetic spectrum over a range of eighty octaves. Only a tiny portion of this spectrum (less than one percent) is perceptible to our "normal" sensory perception. We infer the existence of the remainder from careful observation and the reading of instruments. By extending our senses mechanically, we have become aware of a universe full of subtle, invisible vibrations which have known physical effects. Awareness of electricity, magnetism, infrared and ultraviolet radiation, X-rays, etc. has brought increasing control and use to fundamentally change the world in which we live.

---

* Consciousness researcher and explorer John White (partially paraphrased above) notes that: "Some scientists -- Sir Russell Brain, for example -- have posited that many things may be learned about consciousness, but consciousness itself cannot be defined on other than a subjective basis or in terms of something less fundamental than consciousness. That is, we may accumulate data on what takes place during consciousness, but never on the actual qualities of consciousness itself." (23) Like a Sphinx, consciousness just sits silently watching.

With the advent of brain research we have begun mapping another, subatomic spectrum of light – thought waves -- which also have known physical effects. (13)* This spectrum begins with the more familiar alpha, beta and delta brain waves and extends into the uncharted reaches of the mind. With biofeedback devices, concentration techniques and various forms of meditation we gain awareness and control of these higher energies in us. But in all our investigation, control, and use of energy, whether physical or mental, we need to remember something. It is consciousness that makes sense of the senses. It is consciousness in which thoughts and dreams arise, subsist, and pass away. With consciousness we invent and construct measuring apparatus, conduct experiments, and analyze and interpret the results. Without consciousness we could not have a concept of what energy is, nor the experience of life. Consciousness, then, seems more fundamental than energy, for all theory about and experience of energy takes place in consciousness. It begins to appear that energy and consciousness are one and that consciousness is the spiritual dimension of the cosmos.

The purpose of this section is to demonstrate how energy and consciousness *are* one. Chapter One briefly explains how energy fields interact to condition and form this body we wear. There is a basic, primordial biological energy or life energy which flows through our body continuously. This energy is now visible through a photographic technique called electrophotography. It reveals to our sight that we are a body of light. Chapter Two examines our body of light and its seven primary bioenergy centers which act like lenses for regulating the flow of this life energy into and through us. We will learn how these seven centers correspond with seven frequency ranges, or octaves, of color vibrations (the rainbow spectrum). Next we will examine their physiological connections with the seven basic nerve plexuses of the central nervous system and the seven major glands of the endocrine hormone system. Finally, we will match these seven centers with their psychological connections -- seven levels of consciousness which create seven time worlds and seven love drives. This will give us a "conceptual map" so that we can consciously participate in some exercises to experience our bioenergetic or light body and to feel these energy centers and their effects for ourself. Theory is tested by actual practice and exercises appear throughout the book for this purpose. In Chapter Three, we will find that this bioenergy or life energy corresponds with a basic substance in physics from which the suns generate light. We will learn how to "light a fire" in ourself, by flowing with the breath of life. In this way, we will see and know for ourself that energy and consciousness are truly one.

* Numbers in parenthesis, for example (13), refer to numbered references in the bibliography at the end of each section.

As we participate in this journey, we will join the scientist with the mystic (yogi) in ourself.

This symbolic union represents the synthesis of apparent dualities or the merging of several seeming polarities – inner and outer, real and ideal, micro and macro, male and female, positive and negative, etc. The true scientist has the same goal as the true mystic – knowing the unknown through union with it. In her classic work, *Mysticism,* Evelyn Underhill describes the mystical experience as a supersensory, transcendent consciousness discovered through meditation. "Mysticism in its pure form is the science of ultimates, the science of union with the absolute." In this sense then, mysticism is the vital personal core of spirituality, activated by the quest to experience higher, more purified and expanded states of consciousness. Among Christians, there is a spiritual experience that is sometimes called "Christ Consciousness." Traditional Buddhists seek nirvana, while Zen Buddhists strive for *satori* or *kensho.* Hindus seek union with Godhead (Brahma) or Krishna Consciousness. For Taoists there is the absolute Tao, and to yogins what they call *samadhi*

or *moksha*. Quakers speak of the inner light, Sufis sing of *fana*, and Hermeticists point to the ALL. For Thomas Merton there was the transcendental unconscious, for Abraham Maslow – the peak experience, for Roberto Assagioli -- the supreme synthesis, for Pierre Teilhard de Chardin – the omega (23), and for Christopher Hills there is PURE CONSCIOUSNESS.

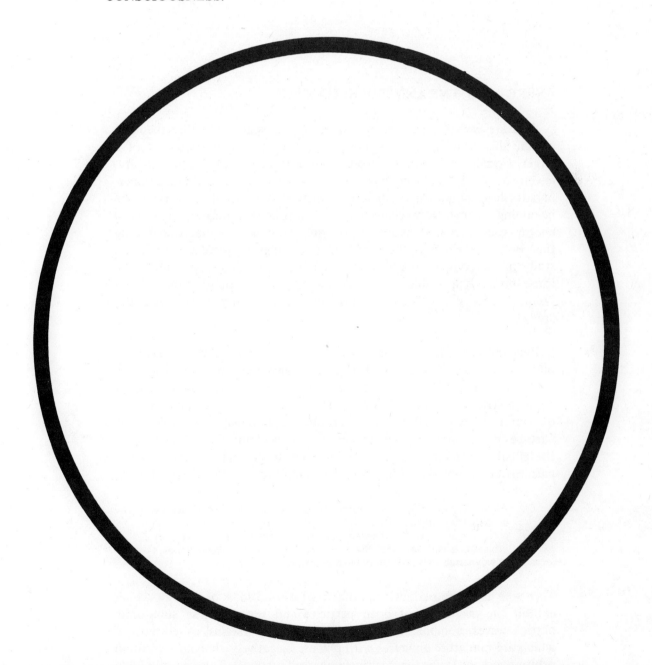

The orb of endless sight, an inward personal sphere of light which radiates outwards into an impersonal creative Void, the absence of all self-consciousness known in the East as Nirvana or Samadhi. In the creative Void the seer does not see anything as being separate from himself, and therefore has no self-consciousness or separated self. This positive self-annihilation from the center within becomes not self-negation but self-fulfillment. ---Christopher Hills, "Nuclear Evolution"

WHO ARE WE?

ENERGY SYSTEMS AND THE HUMAN BODY

Who are we? Are we just bodies occupying space on a tiny speck of cosmic dust, this spaceship-planet we call earth? Or are we part of a larger, Cosmic Whole that surrounds us, interpenetrates us, and finally, envelops us? Do we occupy space or does it occupy us? Philosopher-mystics have pondered the perennial question -- Who am I? -- from the beginning of time. Now science (which is just us thinking) is investigating energy systems and their effects on our minds and bodies and finding that we are much more than we have been taught that we are. From the study of the complex interactions of different energy systems (including those presently invisible, inaudible and nontactile) there is emerging a fascinating story that begins to solve the great mystery of who we really are.

Recent scientific research into the flows of subtle, previously undetected energies within us is opening many minds to some intriguing possibilities. It appears that these bodies we wear are the end product of a whole series of energy transformations taking place in a universal field of vibration, which flows over, around, and *through* us. Subatomic particles (according to some physicists) are *not* tiny bits of matter. At the subatomic level "matter" dissolves or dematerializes into ethereal patterns of energy. According to subatomic physicist Dr. Fritjof Capra:

> All particles can be transmuted into other particles; they can be created from energy and vanish into energy. In this world, classical concepts such as 'elementary particle', 'material substance', or 'isolated object' have lost their meaning; the whole universe appears as a dynamic web of inseparable energy patterns. (5)

This web is composed of spiralling concentrations of energy which exhibit the properties of both particles and waves. These subatomic particle-waves combine into polarized energy systems called atoms, which are miniature universes with planets (negatively charged electrons) orbiting a central sun (the positively charged nucleus). The periodic table of the elements, usually depicted in physics texts as a square-shaped table of linear progression, can be more realistically viewed as circular, a spiralling hierarchy of increasingly complex energy patterns.

This spiralling hierarchy of vibrational polarities applies to molecular structure. The most complex molecule found in nature is the double helix of DNA, the electrochemical messenger of life. Dr. Oliver Reiser suggests that the two chains of the spiral are oppositely charged, thus aiding in the stabilization of the overall molecular structure.

DNA molecules play a key role in the transmission of life codes in the organization of more complex energy systems called cells. Cells, like atoms, are tiny universes with a central sun, the cell nucleus.* Cells combine to form still more complexly organized energy systems which we know very personally and directly as our physical body.

Our body also has a central sun which will be described in Chapters Two and Three. As we learned in Volume I,** our body is also composed of an interconnected system of positive and negative vibrational polarities. We are in turn interconnected with the energy patterns of our spaceship-planet Earth. It too is polarized (on north-south and east-west axes) and has a central sun -- its molten core. Our earth is inseparably linked with the solar system as it spirals like an electron around our sun. Our solar system spirals around the galactic nucleus of the Milky Way. Our galaxy spirals through the entire energy system of the universe which is part of a still larger energy system, the cosmos.

All these energy systems or fields of interconnecting vibrations interpenetrate and influence everything, *including us.* We live in all the worlds here and now and they live in us. We *are* everything, everything is us.

The following illustrations vividly portray the spiral nature of our consciousness. The macrospiral of the cosmos -- the spiral galaxy -- is mirrored in the microspiral of the DNA molecule.

* There is much more space in our bodies than there is "matter." The distance between an atomic nucleus and the electrons orbiting it is equivalent to the distance between the sun and the planets. If all the space in the atoms of our body was removed, the remaining "matter" would be compressed into a ball of energy less than the size of a mustard seed.

** "Dimensions of Electro-Vibratory Phenomena" by Victor R. Beasley, Ph.D., (1975) Volume I in the Supersensitive Life of Man Series, University of the Trees Press.

ENERGY FIELDS AND LIFE

The basic "building block" of our body is the cell. The curious thing about the cell is that it resembles a transceiver (transmitting and receiving device). If we inspect it with a microscope, we notice that the cell walls are folded and convoluted like a semiconductor. In addition, the contents of each cell include organic semiconductors such as liquid crystals, a material which is highly sensitive to minute temperature changes. Most body cells have a double outer membrane. Electrically, this functions like a wet cell battery. Electrochemical processes produce electric currents both within and between our cells (for example, in the nervous system). These currents generate biomagnetic fields, and as Dr. Victor Beasley noted in Volume I: "the individual magnetic fields of all the body's cells and of all the body's systems, combine to yield an overall 'somatic magnetic field,' resulting from all the body's physical, electrochemical, magnetic-producing processes taken collectively."

Soviet scientist Dr. Alexander Dubrov has found that during mitosis (splitting) of a cell there is "an energetic radiation of photons (particle waves of light) which produces a dim glow or bioluminescence."(15) At the same time, he has recorded the radiation of ultrasonic (sound) waves at high frequencies. Dubrov and his colleagues have also observed what they call the "mirror-image effect." Experiments were made with cell cultures in two separate, sealed containers (repeated over 5,000 times). These containers were joined by a quartz window which screened out known radiations, but allowed "eye contact" between the two cultures. After one side was infected with a lethal agent (viruses, chemicals, deadly amounts of radiation) the cells in that box began to die. But so did the completely separated cells on the other side of the window in a mirror-image reflection! To check their theory that the cells might be communicating through some kind of radiation, they used a photomultiplier, which measures the flow of photons. They found that the nature of the photon flow changed noticeably when the cells were infected in one of the boxes.

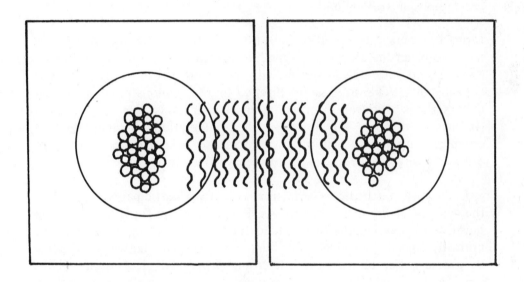

They then theorized that the cells were in some kind of constant communication with one another through coded energy patterns. This idea came from the fact that not only was harm transmitted but also the *identical illness*. Like the fluids in our cells, quartz is a crystalline structure. When plain window glass was used (which is not crystalline) the mirror-image effect did not occur. These observations support Soviet scientist Dr. A.S. Presman's theory of biocommunication.

Presman studied a broad range of effects of the electromagnetic spectrum upon living systems. He found that these systems of energy patterns or fields of wave vibrations "can regulate the spatial orientation of animals, can modulate the rhythm of the organism's physiological processes, and can affect the vital processes of living organisms." His many years of research led him to conclude that:

> Electromagnetic fields normally serve as conveyers of information from the environment to the organism, within the organism, and among organisms.....in the course of evolution organisms have come to employ these fields in conjunction with the well-known sensory, nervous, and endocrine systems in effecting coordination and integration [and adaptation].(20)

His data show that sensitivity to electromagnetic fields increases as we go up the hierarchy of more complex organization. The more complex the living system, the higher the sensitivity. Consciousness researcher Dr. Stanley Krippner comments on Presman's work:

> The fact that sensitivity to electromagnetic fields is greater in fairly complex organized biological systems attests to "organization" as one of the basic principles of life. Thus, a complete understanding of electromagnetic fields will depend more upon the investigation of groups, communities, and populations than individual organisms or parts of organisms.(9)

From this perspective, self-organization is a basic fact of life, if it is taken to mean the ability to receive, store, transcribe, duplicate, generate, and transmit information signals with increasing efficiency, rapidity and dependability. This implies intelligence in the demonstrated ability of living systems to convert information into higher levels of self-organization (to learn) through the process called adaptation. If adaptation is contingent upon information, then increasing sensitivity is required in order to receive the necessary information. How sensitive are we?

Everywhere scientific experiments with us as participants are proving the veracity of Presman's hypothesis. Researchers are finding that we are sensitive to wavelengths that extend far beyond the limits of what we normally consider possible. In Volume I, we learned that we are sensitive to geomagnetic, biomagnetic, electromagnetic and electrostatic fields, ionic and electric currents and many other electrovibratory waves, including thought waves. We heard about the research of Dr. A. Roy Davis who found that, to the site of a broken bone, the body supplies high negative voltage until healing takes place, at which time the normal voltage of that section of the limb is restored. More recent research by Dr. Robert O. Becker correlates with this finding and brings a surprising conclusion. Becker wondered why some amphibians were able to regenerate limbs that had been lost or injured (Remember how chameleons grow new tails?). He found what he calls an "injury current" flowing to the affected part. By using a delicately controlled electric field, he has duplicated this effect mechanically. And he has stimulated previously nonregenerating amphibians and rats into growing new limbs! He and Davis are also reported to be speeding the healing rate of broken bones in people, using this technique. Dr. D.H. Wilson and his colleagues in

Leeds, England, are reported to have speeded the healing of severed nerves in mice by applying a pulsed electromagnetic field.(14) In the Soviet Union, Dr. Alexander Studitsky (at the Institute of Animal Morphology in Moscow) placed some minced up skeletal tissue into a wound in a rat's body. From this the rat was able to grow a new muscle.(14) These discoveries point to the presence of some kind of energy field which acts as an organizing matrix or pregenetic blueprint for our bodily form.

OUR "BIOENERGY BODY"

Soviet electrician and amateur photographer Semyon Kirlian and his wife, Valentina, discovered (as did Nikola Tesla and others before them) a method of photographing the radiation fields and energy patterns emanating from matter. This is a lensless technique for producing images on photographic film or paper using no light source except a luminous or radiant glow on the surface of an object placed in a high voltage, high frequency electric field. Two Americans who reported seeing this effect in the Soviet Union, Sheila Ostrander and Lynn Schroeder, said that the results of photographing a leaf placed in this current,

> revealed a world of myriad dots of energy. Around the edges of the leaf there were turquoise and reddish-yellowish patterns of flares coming out of specific channels of the leaf.... the pattern of luminescence was different for every item, but living things had totally different structural details than normal things.(19)

Deep investigation of this phenomenon is underway. Dr. Thelma Moss and her colleagues at U.C.L.A. have built devices based on the Soviet plans with refinements of their own, as have scientists elsewhere. So far the Moss team has eliminated such physiological variables as galvanic skin response, skin temperature, sweat, and constriction or dilation of blood vessels as causes of this marvel. Their research has shown that changing just one parameter (voltage, pulse, frequency) will dramatically alter the picture. Moss states that:

> There may exist a still-to-be—formulated law of harmonics at work here because the pictures, as we go up the frequency scale from one cycle per second to millions of cycles per second, can be seen to emerge as clear and sharp, become blurred, change characteristics, disappear, and then emerge again with brilliant clarity.(15)

She has verified that inorganic objects have a constant field, and that living things "reveal a fascinatingly varied corona and surface structure depending on the state of the organism (healthy or diseased, aroused or relaxed, etc.)." In short, the "bioenergy body" (as she calls it) is a coherent system, well organized, with a definite shape and configuration for each different organism.

In experiments with people, Moss has found that changes in bioenergy patterns correspond with changes in our thoughts, emotions and states of consciousness. Studies involving close friends or members of the opposite sex usually show a wider and more brilliant glow. In addition, the photographs of two people meditating together typically show a merging and uniting of their bioenergetic fields.

This has led to further studies of people in different states of consciousness. Studies with people in relaxed states produced by meditation, hypnosis, alcohol and drugs generally show a wider and more intense glow. In states of arousal, tension or emotional excitement red blotches appear on the color film. Researchers have no explanation for this, but Moss feels that it may have something to do with "the nonverbal transactions between people which today we describe as empathetic and nonempathetic feeling states." In other words, our thoughts, emotions, likes and dislikes generate energies or vibratory frequencies which have visible results in molding this form we call our body. (In Section III we will see how vibration and form are intimately related.) Our thoughts and feelings, then, influence our body and help create our experience of our world.

A PHANTOM REVEALED

One of the most significant findings emerging from this research is the "phantom effect." Electrophotographic experiments with freshly picked leaves from which a small section of the leaf has been cut away sometimes reveal the bioenergy pattern for the whole leaf! First reported by Soviet scientists, it has been repeated by Moss. It has led her to speculate that this might help to explain a long-standing medical puzzle.

> In the practice of medicine it is not uncommon for a patient to feel intense pain where a limb <u>had been</u>. This phenomenon is called 'phantom limb pain' for which there is no fully satisfactory anatomical explanation.(15)

Dr. Worsley, at the College of Chinese Acupuncture in England, reports that "we can see the shape of a 'lost limb' following amputations. The more pronounced the 'phantom pains' are in the amputee, the more visible is the amputated portion of the body."(9)

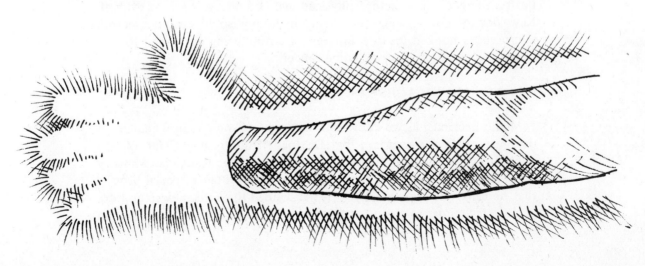

This discovery is of profound importance, when taken with the experiments of Becker in growing new limbs. If the whole bioenergetic field continues to exist after the limb is removed, then *it is entirely possible that the field generates our body,* rather than our body generating the field, *especially when stimulation of the field with energy can produce a new limb.* There is an ancient system of healing based on stimulation of energy flows in our body which still flourishes today -- acupuncture.

Dr. Gaikin, a Leningrad surgeon, studied the Kirlian research and noted that in living organisms there appeared in the photos a distinct pattern of energy points. These points, he found (as did Moss), do not correspond with nerve endings in the skin or other physiological characteristics. Intrigued, he sent the Kirlians some ancient Chinese acupuncture charts of acupoint distribution. To their mutual surprise and delight, they matched! (19) The flares in the Kirlian photos correspond with the acupoints and the channels of energy flow match the meridians.

Soviet scientist Dr. Victor Inyushin has used electrophotography to show a connection between the acupoints on the skin and different parts of our body. After measuring the intensity of luminescence at seven acupoints said by acupuncture theory to be connected with the teeth and mouth, he irradiated the mouths of volunteers with laser light and again measured the luminescence. The intensity of the glow increased by one to two times, while a control area having no acupoints remained the same. He repeated this experiment several times with similar results which led him to declare that: "objective control of the electrical state of the human body can be accomplished by using the acupuncture points and conductance channels and is highly realistic." (18)

Some Soviet scientists claim that pathological conditions show up in the bioenergetic field before physical signs of disease become evident. New devices for viewing this field, including sophisticated electron microscopes hooked up to computers, are now being developed for the early diagnosis of disease, including cancer.

A CANCER DETECTOR?

Medical science, as we know it, is undergoing some radical changes as the result of an invention by Dr. Ioan Dumitrescu, Chief of the Laboratory for Labor Protection and Hygiene in the Rumanian Ministry of Chemical Industry, Budapest. (25) He has invented a special process that he calls "electronography." He has refined the Kirlian electrophotographic technique to the point where "I found I could not only obtain images of bionic energy around the body, but images of the pulses transmitted by individual organs." By hooking the device to a powerful microscope, he was able to capture for the first time the glow of a single cell. He found that healthy tissue appears darker than unhealthy tissue and by this technique he is able to detect and locate "a microscopic-sized tumor months or years before it could be discovered by conventional methods." By connecting his equipment to a computer, he is able to examine five hundred patients a day. In a mass screening of six thousand chemical industry workers, he diagnosed cancer tumors in forty-seven of them. Conventional laboratory tests later confirmed that forty-one had cancer. And it was his opinion that the remaining six would eventually be diagnosed as cancerous when the tumors grew larger. Annual electronographic tests are underway for all Rumanians. Professor Young S. Kim, associate professor of physics at the Ohio State University, feels that this use of our body's own bioenergetic fields in medicine is a major breakthrough. "It provides an entirely new diagnostic tool for doctors. It will be a tremendous help in giving earlier detection of diseased organs, and particularly it gives hope to victims of cancer."

ACUPUNCTURE RESEARCH

In the ancient healing art of acupuncture, now fast becoming a modern science, the Chinese refer to a fundamental concept of a vital life energy -- Qi (pronounced Ki or chee) – which flows through specific channels in our body called meridians. When this energy flows in a balanced way, we are healthy. When the flow is unbalanced, slowed or blocked, we are not healthy. To acupuncturists, then, bodily symptoms of disease indicate that the flow needs to be corrected. Electronic devices have now been invented which locate the acupoints by detecting differences in skin resistance or impedance. (15) (Skilled acupuncturists can locate them with only their sensitive hands.) The curious thing about acupoints (where stimulation is performed) is that they seem to be highly concentrated vortices or spirals of energy. Soviet scientist Victor G. Adamenko reports that from two acupoints hooked into a sensitive electrical circuit there is an output of 50 microvolts, from several acupoints an output of 150 microvolts, and with autosuggestion, a half-volt. (9) This indicates that the acupuncture meridian system may perform a role similar to a transformer in our body's bioenergy circuits. In addition, the acupoints appear to function as a further extension of the pattern of nodal points discovered by the De La Warr researchers mentioned in Volume I. Nodal points are resonance bonds which function to bind us, and all things, to the creative force-field of the universe.

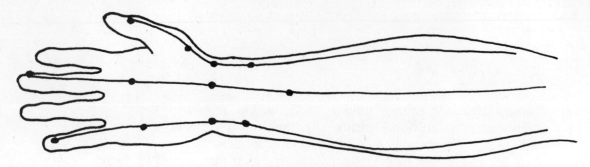

The Yin meridians of the right hand and arm with acupoints. (The Yang meridians run down the back side of the hand and arm.)

Some Soviet scientists have been directing low intensity laser radiation into the acupoints of patients suffering from bone malignancies. Their preliminary work shows reversal of tumor growth and subsequent regeneration of new bone tissue. Drs. Irving Oyle and Jesse S. Wexler of the Headlands Clinic in California have been using ultrasound to stimulate the meridians. (10) They have used this treatment for a wide variety of complaints from back pain to arthritis -- with commendable results. Most patients, they report, showed marked improvement within twenty-four hours and some within thirty minutes. Most patients reported a remarkable relaxation effect and several noticed an increased clarity of vision and intensified color perception after treatment.

Dr. Kim Bong Han, a North Korean scientist, claims to have found that the acupuncture system constitutes a major circulatory system virtually unknown to western physiology. (9) By injecting and tracing radioactive isotope P32, Kim discovered that there are four major acupuncture systems in our body. All four are interconnected through a system of ducts, analogous to the arterioles and venules which tie together the arteries and veins. One system floats free inside, and another is attached to the outside of all the blood and lymph vessels. Another covers the surfaces of and links the internal organs. A fourth is distributed throughout the central and peripheral nervous system. (This would help to explain how severed nerves and blood vessels repair themselves.) Kim and his colleagues have conducted thousands of experiments involving many species to determine the electrical, mechanical, biochemical, anatomical, and embryological characteristics of these meridians and the liquids which flow through them.

Kim says that these systems are submicroscopic and other research shows that the dyes used to stain tissue for microscopic analysis *destroy the meridians* (24); so, naturally, they have long escaped detection. Kim analyzed the fluids and found twice as much adrenaline in the meridians as is found in the blood stream, and at an acupoint, he found ten times as much. This also indicated that the meridians are high-energy pathways, since adrenaline is one of the strongest organic stimulants (injected straight into the hearts of cardiac arrest victims). He reportedly found all the basic cellular components flowing in the meridian fluids. They contain all the essential amino acids, protein and cortical and medullary hormones (indicating a close link with the endocrine system). Nearly twice as much hyaluronic acid is present than is found in sperm in men, and in women there are large amounts of estrogen present, indicating a close link with the reproductive system. Kim also found some unique granules flowing in this fluid which may be the most important discovery of all. These granules have been named Kim Bong Han Sanal. Sanal contains DNA and RNA molecules and protein, and Kim claims to have grown cells from it! During cell division, he reports, the Sanal breaks down into specific molecules known in the West as *chromosomes.*

How does the Sanal get into our cells to form chromosomes? By injecting isotope P32 into a meridian, Kim found that it appeared in cell tissues within forty-eight hours. He performed experiments with mammals, birds, reptiles, amphibia, fish, invertebrates and hydra. He found that meridian systems exist in all multicellular organisms, animal and vegetable, of those he studied. This led him to conclude that the meridian systems are *interconnected with all cell nuclei* in every living system! Experiments with the embryonic chick showed that the meridian systems formed within fifteen hours after conception, at which time the organs are not yet formed. In fact, the organs appear to spring up along the meridians, for they are completed before any of the body parts. The meridians appear to act as some kind of cosmic blueprint or organizing matrix along which the body itself gradually takes shape out of the energy which flows through them to the centers of all our cells. According to Dr. Jeffry Mishlove some of Kim's research has been repeated by Japanese scientists. (14).

These discoveries indicate that our physical body is merely crystallized or trapped energy. We appear to have another body, an energy body, visual proof of which appears in electrophotography. This body is directly affected by our thoughts, emotions, likes and dislikes -- by our levels and states of consciousness. A glow or aura which extends for a few inches from everything shows up in the photos. This points to the existence of yet a much larger aura, one which the mystics have taught about for ages. We will explore this aura in the next chapter.

WE ARE LIGHT

Now we know that our body is a complex energy pattern. Thanks to electrophotography, this wonder is beginning to unfold right in front of our eyes.* Recent evidence shows that our physical body is a mirror-image reflection, or dense energy pattern manifestation, of dynamic processes occurring in our bioenergy field. Since our body cells act as communicating devices, we are constantly receiving and transmitting information coded into energy signals. The center of each one of these miniature crystal sets is linked into the high energy nervous and acupuncture systems. Changes in our bioenergetic field signal the onset of disease before it occurs physically. Stimulation of bioenergy flows, as in acupuncture treatment, cures many ailments in our physical body. Application of pulsed electric currents speeds healing of broken bones and may someday lead to the growing of new organs, thus eliminating the need for dangerous and costly transplant operations. From all this, it appears that our body is not made of "matter" after all. It seems, instead, to be crystallized or trapped light which gives the appearance of solidity at a denser level of vibration perceptible to our physical senses.

Finally, and most important of all, changes in these bioenergy flows and patterns are found to correspond with our changing levels and states of consciousness. Consciousness directly affects the dynamic processes taking place in our bioenergy body, which then cause changes in our physical form. Energy and consciousness interact and flow together. We return to the question posed at the beginning of this chapter -- who are we? We are that universal field of energetic vibrations which manifests in us as our consciousness, our thoughts and our body.

An eclipse of the sun, like a giant eye in the sky.
(Note the similarity to electrophotograph of finger tips.)

---

* Complete instructions for building an electrophotographic device can be found in references 9, 10, and 18.

OUR RAINBOW BODY OF LIGHT

THE ETHERIC AURA

The glow revealed in electrophotography has been described as an aura. Some researchers have jumped to the conclusion that it is *the* aura of mystical lore, symbolized by halos around the heads of spiritually evolved beings in religious art.

However, study of the reports of those with clairvoyant sight (explored in Chapter 5) indicates that the aura they see is more complex and of much larger dimensions (potentially encompassing the infinite). The "Kirlian aura" is found to extend only a few inches from the surface of living things. The etheric* aura, however, is reported to be composed of layers of subtle, normally invisible colored light, and to extend for several feet. This aura encases us in a series of "energy sheaths" which together comprise a sort of "energy bubble." Clairvoyants such as the late Edgar Cayce have long used disturbances in a person's aura to diagnose disease. They also reportedly see indications of which psychological drives and biological energies within a person are deficient and which are in excess. As this potential faculty of expanded seeing manifests in more and more of us as we purify and expand our consciousness, the etheric aura will be meaningless to us unless we have some understanding of what the different radiations mean in terms of psychology and physiology. For this reason, we begin at the center of being, the source of all radiation.

THE CENTER CONCEPT

The center symbolizes the eternal potential.

> From the same inexhaustible source all seeds grow and develop, all cells realize their function; even down to the atom there is none without its nucleus, its sun-seed about which revolve its component particles. As in the atom, so in the stars, -- modern thought only confirms the Hermetic axiom, "As above, so below."
>
> --- José Argüelles, "Mandala"

The center is the beginning, the source, the origin of all vibration, all form, all processes in the manifested universe of creation. It exists continuously, first as seed, then as stem and finally as flower, where a new seed is created. Though infinite in number, all centers are essentially one, for "each is the same irreducible point, the primary syllable, the word, the Logos, through which all is uttered and through which all must pass. This is the significance of the mystical syllable OM, which in its way is the seed center of all sound, as the point is of vision." (1)

---

* Ether is the name given to the all-pervading medium which fills all of space and acts as a carrier for all energy, or light, just as invisible air acts as a carrier for wind (and airplanes). The ether is explored in detail in Volume III, "Supersensonics."

THE SYMBOL OF NUCLEAR EVOLUTION

The still, peaceful center, symbolized by the OM symbol, is surrounded by a deep vortex of changing patterns of spiritual energy. This symbol represents the seer's eye, its iris and its retina, a replication of the Spiritual Sun.

All sound, all vibration, begins and ends in a silent center of zero vibration. The frequency of a vibrating source can only be measured in relationship to a fixed, non-periodic absolute, a timeless domain without frequency. According to Dr. Christopher Hills:

> "One cycle" has only a meaningful relationship to an absolute -- zero cycle; it cannot really have significance to human consciousness unless we say one cycle per . . . and fix the movement in time, in its relationship to zero-time. (6)

The principle of the center remains a concept, a name only, for "an absolute frequency position is not a material point in space but a void created in the mind, a void in the spirit and a void in the ether in exactly the same way our minds create the reality of a meter or foot unit of measure." (6) It is this timeless center which is the beginning and ending point of nuclear evolution -- the domain of consciousness where time is annihilated, and with it the observer consciousness of the separated self.

From this absolute zero frequency position, frequencies beyond the 80 octaves of electromagnetic waves have been measured. This includes 80 octaves of subatomic or etheric energy (corresponding with thought waves), and an additional 80 octaves of superatomic or spiritual energy (corresponding with waves of the Prime Imagination -- Cosmic or Universal Intelligence). (13) This has resulted in the discovery that any vibrating source, including us, sets up resonances in a concentric spherical field with bands of color manifesting according to a particular beat frequency, just as electrons orbit in up to seven bands around the atomic nucleus.

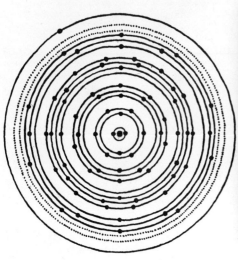  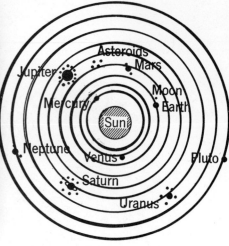

A flat representation of an atom of gold.

Artist's conception of an atom of oxygen.

The Solar System. (These diagrams are not drawn to scale.)

This has enabled consciousness researchers to map the wave fields around people which reveal different drives, levels, or layers of consciousness which correspond to different time domains and colors. Consciousness, as it manifests in us from the timeless center, appears as a spiral of circles within circles and spheres within spheres — a system of hierarchically arranged energy centers or nodal points.

THE CHAKRA ENERGY CENTERS

In our etheric aura we have a series of rotating energy centers which, whether we know it or not, are spinning away inside us right now. Although many of us may not see or feel them, this does not mean they do not exist. We cannot see or feel radio waves although they are all around us and constantly passing through us. All we have to do is turn on a radio and we know they are there.

We cannot see heat though we sometimes see the by-products of its presence such as smoke, flame, steam, etc. But let us touch a hot iron, we sure can feel it! Similarly we can experience these energy centers and later we will find out how to consciously tune in to them. First, to insure fruitful practice, let's see what they are and how they work.

Chakra is a Sanskrit word which means revolving wheel. It is used to represent a spinning or spiralling vortex of energy at a definite location in our etheric aura. There are seven major chakras recognized in Yoga teachings situated near seven corresponding nerve plexuses distributed along the spinal column. They are also positioned in close alignment with different glands in the endocrine system which are responsible for maintaining the dynamic equilibrium or homeostatic balance of the constant environment in our body. As previously reported in Volume I, secretions of the endocrine glands are directly linked to changes in personality characteristics. By changing a single component of our endocrine secretions, we change our personality. The triggering force behind these changes is the flow of "psychic chakra electricity" coming through these energy centers.

Bodily changes are merely physical reflections or "mirror images" of what is occurring on more subtle levels than we normally experience. These levels are governed by energies flowing into and out of us through these vortices of psychic light. Each chakra resonates with a portion of vibratory frequencies from the total energy or light spectrum. They act like lenses, absorbing and radiating energy in the various body areas associated with each chakra. Each chakra controls a certain zone of activity with each level being interconnected with the others. Together they act to filter or crystallize energy from the universal field of vibrations in which we move and have our being. These vibrations are shaped by us into the world we experience, reflected most noticeably in the form of our body. Consequently, the study of energy flow through these lenses is the study of ourself.

An actual clear-seeing (clairvoyant) view of these chakras pulsating with psychic electricity in left and right hand spirals looks like the wheel symbol at left, the number of spokes or petals varying from chakra to chakra.

Teachings concerning these subtle energy centers lie in the heart of many great spiritual traditions. They appear in Christian mysticism, Hinduism, Buddhism, the Hopi Indian tradition, Sufism, and Yoga. There seems to be general agreement on the physiological placement and practical means of experiencing them though there are some philosophical differences. These differences, however, are just varying interpretations of the same basic facts which each school of thought accepts. Just as travelers of different temperaments, languages or interests may describe the same landscape in different ways yet not contradict one another, the same landscapes of human consciousness are described in different ways using different terms and analogies. We will use the Sanskrit system and terms of Yoga here, because they are the most widely recognized and used, the most consistent with scientific investigation and because Yoga is a mystical science of union based upon experiential understanding of this system of centers.

The Yoga tradition sometimes describes the chakras as lotuses and assigns a certain number of petals to each one. Translated into more scientific terms this is merely a way of describing the rate of vibration or frequency range of energy at a particular center. The number of petals in each lotus can be likened to the lines of force that each center radiates. The colors the chakras emit, absorb and resonate with depend on their speed of revolution, which varies with each of us. Colors are just energy vibrating at different frequencies, some visible, some not. Therefore, looking at a color will tend to produce a corresponding vibration to that color in us. Each chakra has a color band associated with it and will be affected by that color. A modern test of human psychological and physiological reactions to color, the Luscher Color Test, is proving very helpful in pinpointing psychological causes of functional disturbances -- dis-ease. This is also true of sound. Each chakra has an octave of sound connected with it and will be affected by it. A special form of chanting known as troupad or free-style resonates the chakras, rhythmically harmonizing them with the Cosmic Sound Current or AUM. This brings experience of expanded states of consciousness.* Like color, the slower vibrations of sound such as low base notes affect the lower chakras, while the higher notes influence the upper ones. Starting with the root chakra at the base of the spine and moving up to the crown chakra at the top of the head, the proportionality of these vibrations is given by the Yoga tradition as: 4, 6, 10, 12, 16, 96, 960. (21)

THE SEVEN SUNS

NO.	GLAND	LOCATION	NERVE PLEXUS
7	pineal	top of head	dendrites in brain
6	pituitary	mid brow	medulla oblongata
5	thyroid	throat	pharyngeal
4	thymus	mid chest	cardiac
3	adrenals	navel	coeliac, solar
2	spleen	lower abdomen	splenic
1	gonads	base of spine	sacral

* Troupad chanting is a central feature of life at the University of the Trees, and special tapes of meditation chants are available from the publishers. Chanting is covered in more detail in Chapter Seven.

THE PHYSIOLOGY OF THE CHAKRAS

The first center is associated with the area of the sacral nerve plexus at the base of the spine. It is known as the root center or Muladhara in Sanskrit (mula means first, adhara means foundation, support). It influences the operation of the sexual organs, or gonads and the operations of the reproductive system of self-replication. It manifests as a desire for union at the physical level -- the mating drive -- and resonates with the color band of *red*. The most pronounced activity of this center manifests at puberty. During this period, the production of sexual hormones, or androgens, results in changes all over the body. (4) The sexual organs develop, pubic and body hair increases, increased muscle protein is laid down, etc. The gonads are capable of generating a tremendous amount of physical energy. This becomes especially evident to those practicing spiritual disciplines, for it is usually the first center to be stimulated by the influx of psychic energy. When this miraculous energy for self-transformation is channelled into creativity other than physical sex, the Cosmic Orgasm of self-transcendence becomes possible.

The second center, the Swadisthana or spleen center, is connected with the area of the splenic nerve plexus and influences the operation of the spleen. However, the spleen acts as a blood purifier and is not an endocrine gland nor is it essential to life. This chakra (as are the other six) is merely located in the area of a nerve plexus rather than having a direct, functional connection with the organ itself. The spleen also influences the breakdown, assimilation and elimination of the life-giving energy in food. The spleen center stimulates the love drive for union at the social level. It resonates with the color band of *orange*.

The third center is a dynamic hub of activity. It is called the solar center or Manipura in Sanskrit and it is connected with the area of the coeliac and solar nerve plexuses. It influences the operation of the adrenal glands which produce a powerful nerve stimulant -- adrenaline. The medulla portion of the adrenals is composed of primitive nerve cells called chromaffin tissue. These cells act both as carriers of nerve impulses (nervous electricity) and secretors of adrenaline. They are found in all the major nerve plexuses of the body. This provides anatomical evidence of dynamic hubs of activity or energy centers at these "power points." The curious thing about these neuroendocrine cells is that they are distributed all along the sympathetic nerve chains running along each side of the spine. *This distribution follows the same basic spiral pattern as the double helix found in the DNA molecule.* (The importance of these facts will become more evident in the next chapter, where energy-flows in our etheric field will be explored in more detail). This chakra resonates with the color band of *yellow*. It represents the drive for or love of change expressed as a desire for freedom of self-assertiveness and self-expression.

In the middle of our chest is the fourth center connected with the area of the cardiac nerve plexus. This is the Anahata or heart center. It influences the operation of the thymus gland which is responsible for the complex system of immunization through the lymphocytes and leucocytes. It has been linked with the response of the adrenals to stress situations and appears to act as a mediator. This corresponds with the basic drive for self-security with which this chakra is concerned. It resonates with the color band of *green*.

The next area is the site of the fifth chakra, the Vishuddha or throat center. This chakra is connected with the area of the pharyngeal plexus and regulates the thyroid and parathyroid glands. The primary hormone produced here is called thyroxine. Thyroxine regulates the general metabolic rate of the entire body. It also regulates oxidation processes occurring via the blood, since it is carried in the blood plasma. Consequently, it is responsible for regulating our intake of energy from the food we eat and the air we breathe, as well as brain cell oxidation and metabolism. The throat center is the linkage point between the three chakras related to mental activity (imagination, intuition and conceptualization) and the four lower frequency chakras which regulate physical and emotional activity. It commands expression and communication. The timbre and quality of the voice indicate the degree to which this chakra is developed. It represents the drive for authority and resonates with the color band of *blue*.

In our forehead above and behind and between the eyes is the sixth, Ajna or brow center. In mystical lore this center is sometimes called the Third Eye, the mind's eye, the organ of inner vision or the orb of endless sight. It is associated with the area of the medulla oblongata portion of the nervous system at the base of the brain. It influences the operations of the pituitary gland which is also called the master controller of the endocrine system. Through its secretions the pituitary regulates the other endocrine glands. It is tied to growth or lack of it, for imbalance in pituitary hormones leads to giantism and dwarfism. This is the seat of individual consciousness. Fully opened and expanded this chakra represents the essence of universal mind. By developing the power of meditative concentration we are given greater clarity, effectiveness and decisiveness of mind. This in turn affects our entire body through the more balanced operation of the endocrine system. This chakra resonates with the color *indigo* and represents the drive for an aesthetic supersensitivity to supramental "guiding fields." (6)

At the top of the head is the seventh or crown center, the Sahasrara (or thousand petalled lotus). It is connected with the firing of nerve dendrites of the brain and influences the operation of the pineal gland which contains retinal tissue and acts as a third, though atrophied, photoreceptor. That is, it reacts to light, especially that entering through the eyes (more about this below). Usually inactive in most of us, some mystics have long claimed that this gland is activated when the higher states of consciousness are reached. This chakra resonates with the frequency band of *violet* in the visual spectrum of light. It is sometimes called the void (sunyata) center, and corresponds with the drive for an enduring order which is found only in the guiding logos or divine plan which lies beyond the limits of the imagination.*

HEALTH AND EYE-LIGHT

Pioneering research on the effects of natural and artificial light on plants and animals is underway at the Environmental Health and Light Research Institute in Sarasota, Florida. (17) This institute was founded by John Ott, formerly a time-lapse photographer for Walt Disney Studios and now a full-time photobiologist. He feels that experimental evidence proves that the quality of light *entering the eye* is directly related to the quality of health in us and other animals. He has discovered that the pigment epithelial cells in the human retina (the cells that give our eyes their color) are light sensitive and are apparently connected with the *endocrine system* via the *pineal* and *pituitary* glands. This light-transmitting pathway into our body is especially sensitive to long wavelength ultraviolet light.** In an experiment a person was fitted with an ultraviolet transmitting contact lens for one eye and a nonultraviolet transmitting lens over the other eye.

> Indoors, under artificial light containing no ultraviolet, the size of both pupils appeared the same, but outdoors, under natural sunlight, there was a marked difference. The pupil covered with the ultraviolet transmitting lens was considerably smaller. This would seem to indicate that the photoreceptor mechanism that controls the opening and closing of the iris [the pineal?] responds to ultraviolet wavelengths as well as visible light. (17)

Since the pupil remains smaller than normal when the ultraviolet wavelengths are blocked from entering the eye, the visible part of the spectrum then seems brighter when it is removed. This could explain, Ott feels, why some people feel a greater need for dark glasses.

---

* In Tibetan Buddhism the sixth and seventh centers are joined as one center with dual functions.

** Not to be confused with the dangerous, short wavelength ultraviolet which is mostly screened out by the ozone layer of the earth's atmosphere. Short wavelength ultraviolet is produced by sun lamps and can be especially dangerous to the eyes.

Blocking of the long wavelength ultraviolet (in which life has naturally evolved over millenia) is a standard feature of our artificially lighted environments. Window glass, eyeglasses, windshields, etc., all filter out this essential part of the light spectrum and keep it from reaching our eyes, and our endocrine system. In addition, artificial lighting normally produces *almost no* long wavelength ultraviolet. Most artificial lighting, in fact, is nowhere near the full spectrum of sunlight. Incandescents produce light in the red and infrared part of the spectrum. Fluorescents produce light in different parts of the spectrum depending on the design, but usually are strongly in the yellow area. Finally, and perhaps most significant of all, long wavelength ultraviolet coming to us from the sun is reduced by air pollution.

> Scientists at the Smithsonian Institution in Washington, D.C., report a loss there of 14 percent in the overall intensity of sunlight during the last sixty years. Scientists at the observatory on Mount Wilson in California report not only a loss of 10 percent during the last fifty years in the average intensity of sunlight at that high elevation, but a 26 percent reduction in the ultraviolet part of the spectrum. (17)

These startling revelations indicate that many of us may be suffering a new dis-ease, another by-product of the industrial age -- light deficiency.

Ott reports an experiment with cancer patients who spent as much time as possible outside in the sunlight without glasses (especially sunglasses) and avoided artificial light sources (including television). All but one patient showed some kind of improvement, and that person continued to wear her glasses. He also relates the story of a school in Niles, Illinois, which in 1961 had the highest rate of leukemia cases of any school in the country. Upon visiting the school he learned that "all of the children who developed leukemia had been located in two classrooms, and that the teachers in these classrooms customarily kept the large curtains drawn at all times across the windows." It was therefore necessary to keep the fluorescent lights on during classes. These lights were found to be very strong in the orange-pink part of the spectrum. The curtains were opened, new lights were installed and used only when needed. Upon a subsequent visit four years later he learned that "no explanation for the previous unusually high rate of leukemia at this school has ever been found, but the problem no longer existed and the situation had returned to normal."

Ott also makes some convincing correlations between the cancer-producing effects of certain chemicals in laboratory animals in combination with different kinds of artificial light environments. This raises the interesting possibility that artificial lighting is just as important in the incidence of cancer as the so-called carcinogens (cancer causing agents). Cell biologist Dr. Matthews O. Bradley and his technical assistant Nancy O. Sharkey of the National Cancer Institute's Laboratory of Molecular

Pharmacology in Bethesda, Maryland, have issued an interesting report. (26) They found that the light from fluorescent lamps can cause mutations in the chromosomes of hamster cells grown in glass dishes. These mutations may be partially responsible for transforming normal cells into cancer cells. They plan to do experiments with human cells to see if they are damaged in the same way. From their findings they conclude that previous research on mutations in animal cells may have been confused by the effect of fluorescent light.

If this is confirmed through further investigation, then it will have far-reaching implications for the entire industrialized world. The chemicals in the food we eat, the water we drink and the air we breathe combined with the light we ingest through our eyes, our body and our chakra energy center system make us walking laboratories. What are we doing to ourselves?

> In 1950, the United States had the fifth lowest infant mortality rate in the world, but eighteen years later we had dropped to thirteenth place. . . . The U.S. ranks thirty-seventh among nations as to the life expectancy of twenty-year-old men, and twenty-second for women of the same age. Something is obviously causing an alarming deterioration of the national health record of this country, and this may lead to disaster if the trend is not reversed. (17)

New inventions may help us out, if we use them. The Vita-Lite fluorescent bulb is specially designed to emit light which corresponds very closely with the normal sunlight spectrum though with less intensity, of course. Ultraviolet transmitting eyeglasses and grey-tinted sunglasses are on the market. Ultraviolet transmitting plastic window panes are available. There are also "Black-light" (long wavelength ultraviolet) fluorescent lights available which can be used occasionally to counteract the increasing absence of this biologically important energy in our increasingly artificial light environment. Finally, there is sunlight, which is free.

When sunlight passes through a prism, such as raindrops falling from the sky, it splits into the seven colors of the rainbow spectrum. Secondary light, reflected into our eyes and experienced as color, has definite psychological effects. The seven colors of the rainbow spectrum correspond with seven levels of consciousness in the human rainbow of the etheric aura.

SEVEN COLORS AND SEVEN LEVELS OF CONSCIOUSNESS

A color, as we perceive it, has no reality by itself. A wavelength of light can be measured and defined as light of a certain frequency of vibration. For example, light with a wavelength of 650 millimicrons is defined as red light. But this in no way guarantees that it will be perceived as red light. The existence of *visual color depends upon a standard for comparison.* Frequently this is a memorized standard, but usually it is based upon a comparison with surrounding colors. This comparison is automatic and influences what we think we see. Light from a tungsten bulb, for instance, looks white indoors, yet in sunlight it appears distinctly blue or violet. The experiments of Edwin Land with simultaneous color contrast expose the fallacy of assuming that our psychophysiological experience of color corresponds with specific frequencies of light. Land showed that all the major colors of the spectrum can be found in a frequency range of only 15 to 20 millimicrons of yellow light, by simply making maximum use of the creation of color in one area by juxtaposition to its opposite or complementary color. Color therefore is relative. It is a system of proportional relations rather than fixed frequency bands. Color is processed *dynamically* by our psychophysical visual system, not statically like a camera or other mechanical instrument would process it -- as is still believed and taught in many places. Thus the very concept of color is devoid of meaning except by including the consciousness of the perceiver in the concept.

The perception of any visual image involves three main stages. First, photons (light) impinge on the retina where they are converted into electrical nerve impulses. Next this electricity is transmitted to the brain through a progressive series of electrochemical reactions in the nerve cell chains. Finally the occipital lobes of our brain receive this electricity and convert it into the images that we think we see lit up outside us when they are really inside us, inside our consciousness. Thus, certain psychoactive drugs (LSD for example) are notorious for inducing the perception of visual images which do not have any apparent external origin. These are called hallucinations by those who do not see the same "external" world as the drugtaker, though the images are frighteningly or ecstatically real enough for the person perceiving them. Drugs act on the perceiver of the images, not on the images themselves. Yet altering consciousness alters perception and what is real is therefore relative to the state of consciousness of the perceiver. We are our perception – the perceiver, the perceived and the entire process of perceiving. (More on this in Chapters Four and Five.)

There are several new branches of science or psychosciences devoted to the study of the complex interactions of thoughts, emotions and body electrochemistry which spring from and lead to altered states of consciousness. Psychochemistry and psychoneurology study mind drugs produced by our body and by the new science of psychopharmacology and their psychological and physiological effects on consciousness. Unfortunately, in many cases experiments are performed without the consent or sometimes even the knowledge of the participants. Behavior modification through the use of psychochemicals in schools, mental hospitals and prisons is really behavior control and the psychochemists and psychopharmacologists who produce these drugs bear some of the responsibility for the uses to which they are put. They need to study the findings of another new science -- psychoendocrinology -- in the effects of color and music on personality via the hormonal secretions of the endocrine glands. Psychoendocrinologists and other psychoscientists need to study the findings of psychophysiology concerning the chakra energy centers and polarity therapy of psychomedicine. All of us need to study the effects of color and sound (music) on human consciousness -- our own -- through experiential exercises.

The levels of consciousness are like universes that interconnect and coexist, concentrically, within each other (like skins of an onion). Each operates within a certain zone of activity and rarely do they come into direct contact and share the same experiences (except through the syntheses of nuclear evolution). Each color can be linked with a level of consciousness, but most people operate in several. Our whole being is essentially a huge synthesis of complementary opposites, an organic unity which continuously grows and changes.

In order to see these levels of consciousness manifesting, we must take a momentary glimpse or have an insight into a person's life. It is the consciousness of the moment that determines our drives, conflicts, and actions. A person may in fact live predominantly in one particular level of consciousness. But the way in which any level manifests depends on the whole being, for in isolation a level of consciousness has little meaning. In our life situation we have several levels in operation, each in varying degrees of ebb and flow. Here the levels come to have real meaning for they are no longer just descriptions, they provide explanations because we can see them operating. Therefore, a certain action can be described as the creation of a, for example, red-level consciousness. But this explanation depends on getting into the inner world of the person and seeing how the basic energies interact. This is what makes the explanation specific, for color is only meaningful in relation to us as whole beings. Red, for example, can mean passion, aggressiveness, excitation, or anger. Which it means depends on many other factors which only empathy can reveal.

Therefore color is directly related to psychological states or levels of consciousness which have their corresponding emotional or love drives. Each chakra is linked with a love drive or desire for union associated with that level. For instance, an emotional change from calmness to rage is triggered by thoughts. This results in certain perceivable physiological changes such as increased flow of adrenaline, constricted blood vessels and higher blood pressure, faster breathing and heart beat, muscular tension in the abdomen, etc. These changes are triggered by an increased flow of endocrine secretions from the adrenals and thyroid glands. Particularly important here also is the role of the pituitary because it releases hormones which stimulate or tranquilize the adrenals and thyroids. The action of the pituitary is closely linked with our thoughts and the energy streaming in and out through the chakras located in the head. Its secretions control our physical expression of emotion by way of our thoughts. This confirms the psychological principle that energy follows thought. Consequently, the links between the seven chakras, the endocrine glands, the seven colors and the seven levels of consciousness are vital and fundamental ones.

For all the levels described below the unifying factor is the experience of time. But the specific way we express this in our personality depends on previous experience and the relative proportion of all the time worlds present. The time world is vitally important. It is the experiential variable that is equivalent to the time dimension in color vibrations. Colored photons of light are all the same. It is only the time-space between individual pulses that determines what we experience as a specific frequency which our visual apparatus translates inside our head into a colored form "outside." The experience of color therefore depends on time and frequency. The experience of the different levels of consciousness, then, depends upon the direct experience of the different inner psychological time worlds. According to Christopher Hills, our chakra system is the medium for the experience of time, each of us having an innate tendency to experience time through one particular chakra more than the others. Each of us lives in a time world which is conditioned by the rate of vibration of our most active chakra.*

RED PHYSICAL LEVEL

The time world of the red level of consciousness is the immediate NOW. Past is forgotten and the future is unimportant, for all that is experienced are the many physical stimulations present in the moment. Because the past is forgotten, it is difficult to learn from experience except by physical trial and error or rote learning. Because the future is not immediately important, activity has no need of direction for it has nowhere to be directed. In this level there is no need for planning or foresight, only a need for a continual reaction to the environment. Because action in the immediate present is valued, any action taken now is preferred to delayed action due to planning or questioning. Trial and error, then, is preferred to education by hindsight or action with foresight.

This time world creates a drive for stimulation and excitement all the time. The drive is one for continual manifestation, to physically enact the inner desires. There is a need to be in control of the physical world, a control akin to domination. The love drive for union at the physical level makes sex the most frequent creative outlet for this energy. In fulfilling the need for self-abandonment, we create a physical situation which brings excitement and stimulation. It is a situation experienced entirely in the immediate moment and is tangible. It fulfills the need to experience contact with the universe through the senses by touching the

---

* The following color meanings are those researched by Christopher Hills in "Nuclear Evolution" published by Centre House Publications, London, 1968; newly revised edition University of the Trees Press, 1977.

universe. The sexual partner is tangible and can be experienced skin to skin. The physical viewpoint of this level of consciousness, however, is only one of many ways of experiencing love.

This level of consciousness can also be seen in the drive to change society immediately by destroying that which is not satisfactory. However, the red level revolutionary would rather act and destroy than first discover a new social system that can replace the old one. This level of consciousness which seeks to dominate the environment does not allow for other ways of being included in the social network except within the roles of domination and submission. This can be positive, for such a drive causes effort to be made to actually manifest what is required. However, it will need the consciousness of different levels to know what is really required for successful action.

ORANGE SOCIAL LEVEL

The time world of the orange level extends farther than the red into the future (extends the effects of action taken now). At this orange level, the past is not important in relation to the future. Thus, action is taken now to change the future. This is the time world, for example, of the politician who quickly forgets past inconsistencies and believes he can keep changing the present to improve the future. But because the present is ever-changing, yesterday's future never comes. Promises are easily made and more easily broken. Awareness is limited to social factors which leads to ambition or the creation of a social image.

This time world creates a drive for social contact, for the whole of society to experience what the individual experiences. This is achieved politically by changing the system of social government to suit the politician's beliefs. There are many ways of expressing this drive directly through a social life where contact with others is more direct. Part of the difficulty for a politician, and a cause of ineffectiveness, is the inability to work with people as they really are. By tying individual behavior to social factors the politician tends to create an image of society that does not always include real people. Thus, political systems are always in danger of being broken up by those who disagree, or simply by neglect. An understanding of the need for individuality would allow more freedom for expectations to change and tolerance for natural human lack of awareness.

The drive to dispel self-doubt by comparing one's self-image with others can also create an element of pride. While success is often sought in order to eliminate doubt, it can also be used to create separation between the individual and others in society in the form of pride (a superior versus inferiors). Orange is in between the sensation orientated red level and thought orientated yellow band of consciousness. It is a combination of the need for contact and stimulation and the need for more expansive experience. These needs are fulfilled within a social framework.

YELLOW INTELLECTUAL LEVEL

The time world of the yellow level of consciousness includes the past, present and future. Events are perceived as progressing through time, each event preceding one and following another. This is an awareness of time based on logic. It is an objective evaluation of the manner in which events appear to change. The conception of cause and effect is in operation here, for cause and effect is basically a way of looking at change over a period of time. This level of consciousness is also analytical. To arrange its experience sequentially requires the ordering of memory, present experience and future anticipation as though units of experience could be created out of the total experience. This is achieved by analysis of all that occurs. A person of this level feels nothing can be known unless it has a beginning or background and each additional piece or unit follows logically from the next. It is in this logical development that new thoughts are presumed to develop. However, it is impossible to really understand or to fully know anything by dissecting it into its component parts.

The best example of this level of consciousness is the scientist who tries to logically arrange all the pieces of information, and gains comparative knowledge by abstracting one variable from the total and experimenting with this one factor by holding all others constant. The problem is that we cannot know the whole by looking at the parts; there are too many variables. And the biggest variable of all is the consciousness of the observer. The Gestalt dictum that 'the whole is more than the sum of its parts' expresses clearly why the analyst cannot capture the spirit of life. Thus the desire for an original thought is often frustrated by the limitations of logic, which is sequential and analytical.

The same analytical perception applies to all experience on this level. Analysis automatically creates a separation, an ego that can be distinguished from other egos. Often this occurs as a basic distinction between me or I, and other "me's." Each is perceived as being separate. This means that all perception is objectified or made external. This is really projection for it is impossible to perceive anything without a perceiver. It is the perceiver who determines what is perceived, so it is not possible to know anything without knowing the perceiver. What we see is in ourself. Therefore, the intellect can never prove beyond (logical) doubt that anything is ultimately true. To the intellectual, reality is relative, for one cannot understand the absolute by analysis without dividing it and making it relative.

The effect of this kind of thinking on human relationships is important. The separating intellect divides feelings from thoughts, and the presumed superiority of logical thought can create feelings of self-superiority and separateness. Positively used, it can coordinate the wildest dreams of the poet with the impetuousness of the activist, and produce a plan that brings method to all the other levels of expression.

GREEN SECURITY LEVEL

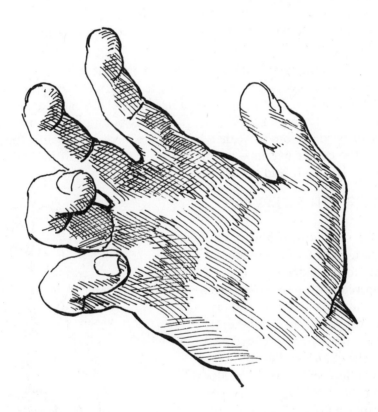

The time world of the green level of consciousness includes past, present and future as does the yellow, but they are not sequential. Rather, the past can follow the future as easily as the present can follow the past. Changes occur depending on the fulfillment of the basic need for stability or balance. This need may be in the form of security (financial, material or emotional), recognition, confirmation or self-respect. In whatever way the balance is attained for that individual at that time, the balance is experienced as a feeling of well being. Where the red level seeks physical excitement, the orange social contact and the yellow mental stimulation, the green level may include all of these if they add to the feeling of security. Of themselves they are not fulfilling. But if they add to the self-respect sought by this level, then they become temporarily fulfilling. The time dimension in which the attention is focused depends upon where the fulfillment of the basic need resides. If a person seeks financial security, then the time world will change according to which time world will have most influence upon that need. If the influence is either to increase or diminish the power to satisfy this need, then the attention will immediately go to that inner time world. Thus at a time of financial crisis this person will be in the future, fearing an eventual loss of security. If the security is lost it is recreated in the memory, the past, while every effort is made in the present to restore the security. Thus a negative, grasping green level consciousness prefers a decreasing loss to decreasing profit, because it brings security closer. Decreasing profit threatens security. Yet the first, which is preferred, is a loss. The second is still a profit, but is feared.

This need for the power of a self-sustaining and self-fulfilling life force is present in most people. Whether or not it is fulfilled depends on the source of the power. If it is ephemeral like financial wealth or depends on people for a feeling of well being, then it is bound by circumstances to be shortlived. If on the other hand the source of the life force is eternal -- the center within -- we do not need to collect it like material things for it is already and always here.

BLUE CONCEPTUAL LEVEL

The time world of the blue level is in the past. In order for the present or future to be understood it must be compared with past experience. This is most common in old age where nostalgia and sentimental memories are often preferred to looking ahead, where "death" (the unknown) lies in wait. It is also prevalent among those who are conservative and cling to traditional ways of doing things because these are the ways they have always done them. The source of much conflict in society is based on the need to have past precedents before accepting a new thing. This is especially true where any change is considered an improvement. By constantly referring back to the past a person experiencing the blue level is only

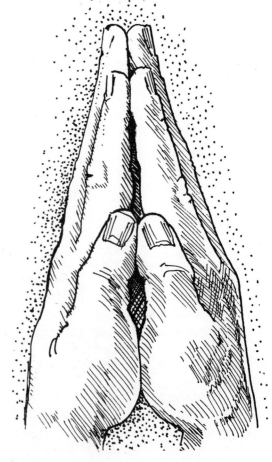

able to accept the new by integrating it with past experience. Any event that cannot be synthesized with the past order of life is immediately rejected. The rejection can come in many forms. To parents who have great attachment to the concept "family," the young who seek to assert their individuality are experienced as breaking up the family. Some parents cannot comprehend why "their" children seek to take the risks involved in independence and to reject their blue level ideals.

Some religious authorities take the same attitude toward those of the faith who disrupt the emotional satisfaction of others of the faith by creating new faiths or demanding change. The rejection of those who disregard the "authority" of the faith can be total and vehement. Yet it is the authorities who are dogmatically inflexible, not the originator of the faith. A similar phenomenon happens in science where new theories or new concepts are only accepted if they conform to the scientific authority established by precedents; no matter that theories, models, paradigms, etc. have come and gone in the past. Completely revolutionary breakthroughs in science happen rarely, for those who achieve these insights need both the imagination to perceive them and the knowledge of science to communicate them.

The drive of this level is to synthesize, to bind events or knowledge together. In science it creates the philosopher or the conceptual thinker who is able to grasp all the salient points of the analysts and bind them together. The philosopher wishes to concentrate on the overarching principles, while the intellectual (yellow level) concentrates on details. Both are important in developing knowledge and communicating it.

INDIGO INTUITIVE LEVEL

The time world of the indigo level of consciousness is in the future, for the basic drive is to go beyond the immediate experience of the moment, to integrate with what is beyond the limited concept of self, to directly perceive what is not yet present in time or space. The time world of this level cannot be comprehended by the previous ones for the reality of the future to this level is perceived as vagueness and abstractness by others. This level experiences the present equally as vague for the present does not exist in awareness except as a memory. In relating each new event to the future and in looking for its future consequence the experience of the moment is lost. It is relegated to a memory. The past, of course, is even more hazy for last week happened so very long ago. In a sense this means that time itself is no longer important, for future time extends infinitely. This makes it harder for those on other levels to understand the lack of concrete action at this level. The imaginative consciousness of the violet band cannot understand why the indigo imagination is so free and breezy. The future for the indigo level is imagined not with specific images but with abstract impressions, where meaning is taken out of context and becomes absolute in itself. Thus there are no meaningful images or concepts for others to grasp.

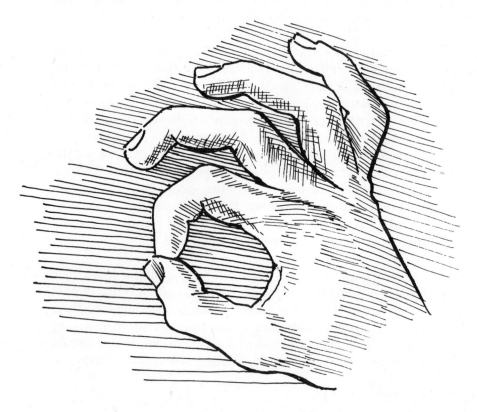

In the indigo level the individual gains receptivity toward the future, toward what cannot be sensed, and the supersensitivity of direct perception is developed (for example, in telepaths and clairvoyants). It is from this level that the limitations of the scientific method are realized. For science includes both induction and deduction (or analysis and synthesis), but both are only ways of *knowing about* as exemplified by the analyst (yellow), the synthesizer (blue) and by the creator of an abstract thesis (indigo).

The frustration of this level is that the future takes so long in appearing. Even the immediate present has little to offer for it is experienced by looking back into the past from the future. But by looking directly to the future the seer can see where those limited by sentient knowledge cannot.

VIOLET IMAGINATIVE LEVEL

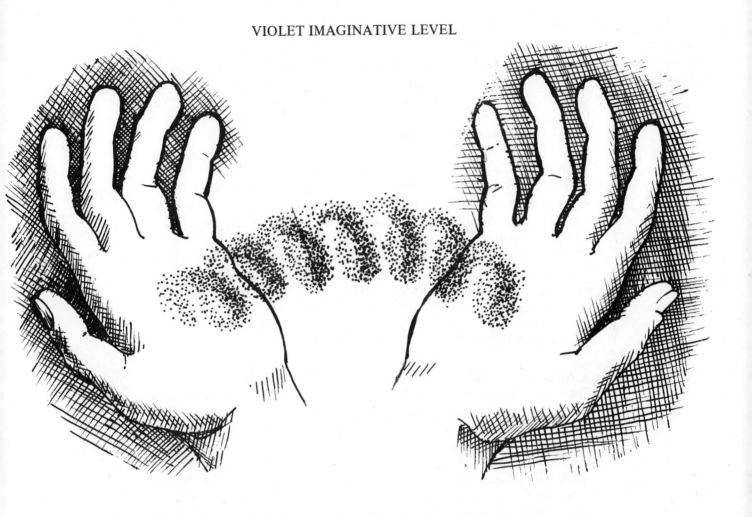

The time world of the violet level of consciousness is now in the sense that it includes all times (past, present and future) and can experience the endurance of time. Thus the past is now as much as the future is now in this experience of *eternal now*. The continual distortion of chronological time can mean that the present moment is lost to the imagination of other "times." It also means that the meaning of distant times can be captured and brought into the present. The mystics and visionaries have had this ability to encapsulate the future in symbols and visions so that it can be experienced in the now. Poets paint word pictures and other artists sculpt visual images from feelings about and visions of what has been and what will be. The imaginative individual can best communicate with those of the other time worlds, for here creativity is so subjective that it expresses a part of our total experience. The violet level person can experience the other time worlds in the imagination, identify with the experience of others and express this experience in symbols and images that show a natural sensitivity toward others.

However, the drive to become identified with the whole of life means that the aesthetic qualities of its positive expression become frustrated in a world that does not respond to the creative. The imaginer can escape into the imagination, distort time and create a fantasy world of utopian beauty. Using charm, this person gets what he/she needs from others and identifies only with what matches the fantasy.

This desire for a "magical relationship" with friends can be as creative as it can be unreal and fanciful. If the "magic" is integrated with the other time worlds it can serve to catalyze the other ways of knowing whether via thoughts, feelings or senses. In isolation a fantasy world is all that is perceived. This makes it difficult for the fantasizer to integrate into society. But the mastery of the concepts of time and space allows us to break through into the region of the unknown Self. Here chaos becomes ordered and enduring principles can be expressed clearly enough for others to understand and to develop.

WHAT LIES BEYOND?

All colors are aspects of a universal energy -- the light of consciousness. They are all energies with frequencies which vary in relation to a constant (by definition). What must lie beyond relativity is the Absolute. But the Absolute cannot be grasped by the relative ways of knowing of the seven bands of consciousness. Each of these various wavelengths of light experienced as levels of consciousness is merely the life force, or spirit, vibrating at different frequencies. It follows naturally that by changing the rate of vibration of our chakras, we raise ourself to a higher level of being. Through refinement of our consciousness by purification in the fire of life's tests, the dross of our lower self drops away and the pure void of true spiritual consciousness is gradually attained. The void is the unadulterated, undisturbed, "sacred space," the divine Absolute.

> In terms of the various levels of consciousness, this "sacred space" cannot be touched (red), it cannot be explored (orange), it cannot be separated (yellow), it cannot be possessed (green), it cannot be conceptualized or bound in time (blue), it cannot be known by psychic faculties (indigo), it cannot be patterned, imagined, ordered or magically conjured (violet). In short, it interpenetrates and underlies every being, it's action and reaction, and is nothing manifested. Since "no-thing" cannot move within space, it is perfect stillness and the only absolute there is in being, the pure consciousness of "I" itself -- the Absolute "I".
>
> Christopher Hills, "Nuclear Evolution"

EXPERIMENT SECTION FOR CHAPTER TWO

SENSING THE ETHERIC AURA

Many persons have gradually learned to see the subtle emanations and dynamic fields around living things through practice of concentrated attention and meditation. It is merely a question of learning to look in the correct way with the proper focus. Most of us are so used to focusing our eyes directly at the surface of things that we miss the more subtle "surfaces" that lie a short distance away.

STEP ONE: Sit quietly in a semidarkened room. Take a few deep breaths with the eyes closed. Now open them and look closely at your hand in between the fingers and at the sides. Focus slightly behind the hand at a distance of from one to two feet. Now hold your hand against a dark background with only subtle, indirect room lighting to see by and continue looking. By focusing a small distance from the hand yet attending to the area adjacent to it, a whitish grey film or faint cloud of flowing, glistening energy can be seen (but do not expect to see it right away, it takes continued practice).

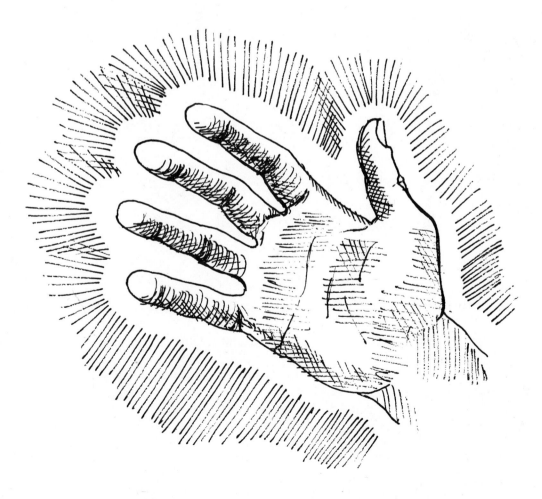

STEP TWO: Rub your hands vigorously together until both hands feel full of energy and become dry. Separate them. In this way a stream of energy can be felt flowing from the right hand to the left. Again look without focusing, see without looking directly. After a while the distance between the hands can be increased considerably and the flow of energy can still be seen.

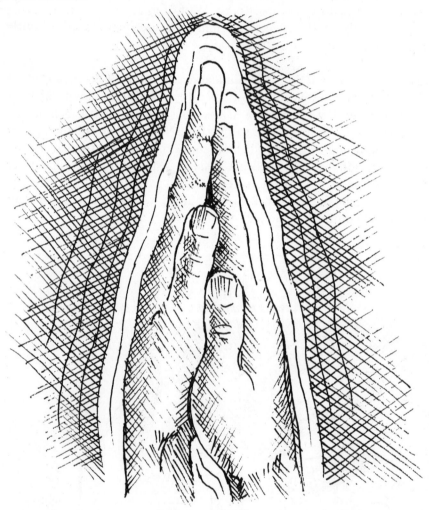

STEP THREE: Next try to see the energy that exists all around a person. It is easiest to see it around someone who is full of vitality (after meditation or deep breathing exercises). Try to look at him or her against a plain background, perceiving the energy field without looking closely for it. This requires a great deal of attention, for the energy is seen directly in the mind, not in the eyes. If your eyes become tired, rest. By practicing seeing with the mind, the auras of everyone will gradually become as visible as the colors of a rainbow in the sky (for those who have trained well).

STEP FOUR: Before falling asleep fill your mind with the suggestion that you are going to perceive your aura. The subconscious mind can be conditioned by the power of suggestion and auras can then become visible in dreams. By continued suggestion this faculty can then be carried over into the waking state.

It is important to understand that the ability to see auras will come at the right time, when you are ready. No inordinate amount of attention should be paid to the auras for they are only one small aspect of energy and life. To perceive only this aspect is to become enslaved by the phenomenon, for auric perception in itself serves no purpose. If you practice to see auras it may take some time. But if you study intensely the perceiver of all phenomena rather than what is perceived, then supersensory perception will naturally occur in a short time.

AN AURA DETECTOR

Auras are difficult to see without the training to see them. Even the dark blue-violet "aura goggles" developed by Dr. Walter Kilner of St. Thomas hospital in London, do not work for everyone. Dr. Kilner discovered a system of diagnosis based on color and consistency changes in the aura by looking through dicyanin stained glass. By looking through these stained-glass goggles, he reports that he can see a cloud of energy around a person which changes with the thoughts, emotions, and (sometimes) will. He found that dis-ease of a psychosomatic origin can be detected in the energy field before it appears in the physical body (confirming the electrophotographic research). These goggles are now being test marketed in some parts of the United States. You can make them yourself by simply staining some glass with dicyanin.

A simple electronic device can be easily constructed that will detect the energy field around the body. This can be done inexpensively by obtaining two neon or argon diodes (light bulbs) and hooking them in parallel with a capacitor resistive network. This system is similar in function to a "flasher unit" in an automobile turn signal indicator. In the case of diodes, however, they are sensitive to electric and ionic changes in the auric energy field. This network sends a pulse to the diode which causes the cathode to heat up and discharge electrons to the anode, creating a flash of light. The anode is connected to the positive side of the system, the cathode to the negative. When the current becomes negative (relative to the positive pulse) the diode does not flash, because the device is constructed so that it pulses in only one direction (see wiring diagram below).

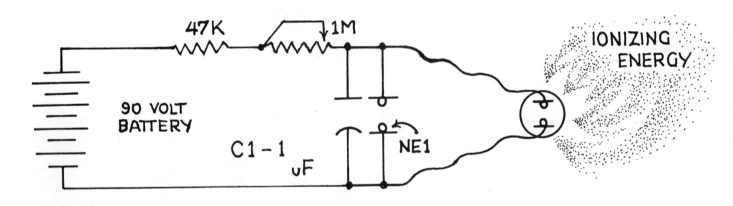

By having two nearly identical diodes in parallel the electrical pulse will light the diode of least resistance, but not the other (unless there is sufficient energy present to light both). This is the basic principle of this aura detector. When the unlit diode is passed through the ionizing field of the aura it will light up, otherwise it will remain unlit.

This detector can show not only the extent of the aura around those energy centers that are radiating energy, but it can also show those that are absorbing it, by ceasing to light up. If the system is made sufficiently sensitive so that small increases of energy are detected, then it is possible to draw a fairly accurate map of the aura and the energy fields around the chakras. It is also possible to consciously create a flow of energy that will light the diode. By placing it between two fingers or close to the palm (but not touching the skin) the diode may or may not immediately light up. If it does not, you can make it do so by concentrating on the diode with a vivid mental image of it lighting up. At the same time relax and breathe deeply while imagining a flow of energy flowing through your fingers or hand. At the moment when concentration is strongest the resulting flash of the diode may be quite unexpected. When concentration is well developed a mere thought carries enough energy to light the diode from several feet away. To keep it pulsing requires tenacious concentration and is best done in stages by slowly moving your hand farther and farther away without letting the attention wander.

SENSING THE CHAKRAS

There are several ways to study the chakras. The simplest way is to sense them in ourself. This is the best way to understand their different roles, by experiencing their operation. This comes about through conditioned careful observation and by controlling their energy and outflow through concentration, suggestion and meditation.

We can sense the energy being absorbed or radiated through the chakras with our hands. By moving a hand toward the body (it is sometimes easier with someone else) we will find a point a few inches from the skin where we feel a sensation. This can usually be felt with both hands, but often the person being sensed prefers the use of either the left or the right. The extent of a person's aura is indicated by how far from the skin this sensation can be felt. By moving your hand over the energy thresholds of the aura in a slow stroking motion, you can find areas of concentrated sensation. These areas may extend farther down the body, or may simply be more powerful. The bottom three centers can sometimes be sensed more easily behind a person, especially if they are wearing heavy clothes. While trying to sense these chakras, both of you should watch any reactions in your body or mind as the hand passes over the various centers (with as much receptivity as possible). If you produce a disturbance at one of the centers, it is best to remove your hand immediately and then try to find the cause of the reaction. You may not find all seven chakras, for they are like doors -- some are locked or only slightly ajar. Not until they are completely open will they manifest their full potentiality.

There are many paths, in Yoga and elsewhere, that lead to the opening of these psychic doors. By following certain Yoga exercises we can develop concentration and arrest the vagaries, trips and distractions of the mind by fixing attention on the chakras. This is similar to an athlete who develops and masters the musculature and poise of the body through appropriate training. Rather than developing physical power, however, we wish to develop the power to know. Seeing and knowing can become independent of the senses and the brain.

THE AURA PENDULUM

Auras and chakra energies can be identified radiesthesically by using the aura pendulum available from the publishers (see Chapter Ten on radiesthesia). This pendulum is a sensitive biofeedback device that rotates when tuned to the colors of a person's chakras. The pendulum merely amplifies the neuromuscular reaction arising from the resonance effect between the color of the chakra and the color indicated on the pendulum. It is an extension of our own perceptive faculties into the invisible part of the light spectrum. The tuning is done by moving a cursor (ring) along the spectrum on the side of the pendulum. In this way, the predominant aura colors can be identified, as well as the condition of the chakras. The aura pendulum can thus be used to discover energies that we are lacking as well as those that are being expressed through us.

> The wooden aura pendulum is black because black absorbs light from the human emanations. By setting the cursor to a color, you can determine what the psychic atmosphere of the person is. There is no need for extra amplifiers in an aura pendulum because the diviner is already unconsciously sensitive to auras. People who are not psychic at all get funny feelings about certain people. (8)

THE SPECTRUM MIRROR PENDULUM

This pendulum is much more sensitive than the aura pendulum described above. It is triply amplified. It has a spectrum along the side with colors selected by moving the cursor, as in the aura pendulum. Its mirror surface reflects off any unwanted waves from the selective power of its spectrum, including negative thought waves. Finally, its hollow core contains a tiny amount of radium ore powder (not harmful) which not only amplifies the colors of the spectrum, but also fires energy out the pointed end, connecting it with the Earth's magnetic field. It thus becomes a greatly extended and vastly powerful pendulum which is highly selective to vertical waves because of its vertical shape (which is the third amplifier). It can be used for balancing the energies of the different chakras by injecting extra or removing excess energy where needed, and to tune in to the causes of blockages and disturbances of energy flow in the aura and remove them, thus balancing the entire etheric aura.

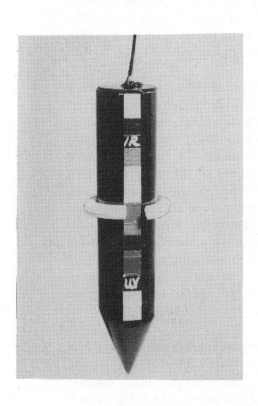
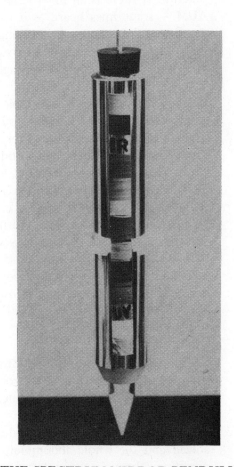

THE AURA PENDULUM   THE SPECTRUM MIRROR PENDULUM

However, in order to use it and other divining tools properly, you will need to read and practice the proper techniques. These are set out in simple step by step instructions in an excellent layman's handbook of divining, *Alive to the Universe* by Robert Massy, Ph.D. (Volume VI in this series). For a deeper investigation into the nature of nature and the universal field see Volume III, *Supersensonics*.

CONCENTRATION

Concentration is an extremely important fundamental step in sensing the chakras. By training our gaze (called tratakam in Yoga) we learn to control the restlessness of the mind. The nasal gaze is a good example. In this exercise we open our eyes half way and stare steadily a short distance beyond the tip of the nose for about five minutes at a time. If the eyes become tired or strained, we merely close the eyes and relax by breathing slowly and deeply. When they recover we continue the practice. If there is an inner reaction we meditate on the inner sound – AUM – sometimes heard as a ringing or buzzing in the ears or nerve dendrites of the brain. This gazing exercise practiced regularly sensitizes the olfactory nerve centers, helps to unfocus the eyes for aura gazing, increases concentration and strengthens memory.

Procedure: Sit in a comfortable position with spine erect and back straight (but not rigid). Close your eyes and place your attention in the psychic center between the eyebrows. With each inhalation of the breath draw your awareness up your body and concentrate it in this one spot. Next, pass a metal object back and forth across the forehead a short distance from the skin. You may be surprised by the experience of a sensation as the object passes. In fact, if you concentrate deeply, you will perceive a dark indigo field in your mind's eye and will be able to see the object as a darker mass against this indigo field. This is your rudimentary Third Eye beginning to awaken. It is always there, waiting to confer on you a type of perception and knowing that is direct, where distance and normal sensory knowing are irrelevant. This is the intuitive part of our mind which *The Supersensitive Life of Man* series is designed to help us develop.

The effects of concentrated attention on each chakra can be beneficial in many ways. Attention at the root center brings control of the powerful sexual energy which can be released up the spine for other creative purposes besides sex. Attention at the spleen center can bring control and magnetism to the legs. Abdominal and metabolic problems can be solved and the whole abdominal area can be filled with strong, healthy vibrations through concentration at the solar center. Concentration at the heart center brings feelings of energy radiating outwards in all directions, a sense of security gradually flooding the entire body with healing positive green energy. The throat center can magnetize the neck and arms. Attention at the brow center brings a feeling of unity and a type of knowing that is direct. Finally, complete attention at the crown center can bring in a flow of energy from higher levels of consciousness which will flood the entire body with bliss, like a star, radiating its life-giving light in all directions.

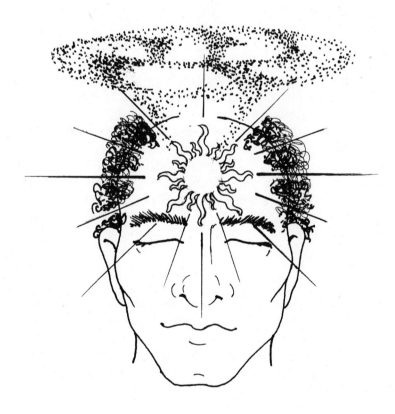

HERE COMES THE SUN

LIFE ENERGY

There is something we do all the time, yet seldom are fully aware of it. We breathe. Without this "umbilical cord" with the cosmos which is the breath life ceases. Without oxygen our physical body cannot continue to operate, to live, to be. Breath *is* life. But what is breath, really?

Breath is far more than the ingestion of oxygen and the excretion of carbon dioxide. Everytime we breathe, we inhale energy. This energy flows to all the cells of our body. It is the fire of the life force which moves all things. It is the primal energy behind all form, from which all energies originate and to which they all return. This energy has many names. We will concentrate on three of them here. Two are ancient and come from the mystical science of Yoga and the healing art of acupuncture. The other is modern and comes from the science of physics. All three appear to embody the same principle – *life energy*. They all represent similar perspectives or viewpoints. The two ancient ones are essentially mystical, experiential and psychological. The modern one is basically scientific, experimental and physical. The veils between inner and outer disappear as we explore these three concepts and find that they are one.

PRANA

Ancient Yogic texts speak of an invisible energy -- prana -- which fills the universe and makes up the suns. Prana is a Sanskrit word meaning primordial (pra) energy (na). Prana is life energy and all other forms of energy are said to arise from it. At the end of a cycle, they all return into it. Prana is the sum total of all energy in the universe both manifest and potential. A modern day definition by Christopher Hills:

> PRANA: The life force or energy of the cosmos which animates material forms and manifests in the seven vibrations of solar energy. It is also present in breath and the vital gases which act as seven nodal points for material vibrations in the manifest universe. PRANA can be compared to the concept of the "Breath of Life" in Genesis activating the created world and the physical being of man.*

---

* From Section Zero of the three-year course to direct enlightenment -- "Into Meditation Now."

Prana represents the total movement of cosmic force or life energy in and through us. We feel its radiant presence in the warmth and tingle that the sun's radiation gives to our skin. We can feel it, too, as a breath of pure energy.

QI OR CHI

The Chinese culture has had a nearly identical concept of life energy which has been the basis of Chinese medicine for several thousand years. With the tremendous revolution occurring now in the field of acupuncture research, one of the fundamental concepts of Chinese thought and a great cross-cultural influence at this time -- Qi or Chi -- needs a deeper examination. A modern day definition by Dr. Felix Mann (this and the following extracted from *Acupuncture: Cure of Many Diseases*):

> The manifestation of any invisible force, whether it be the growth of a plant, the movement of an arm or the deafening thunder of a storm, is called Qi: though...there are many varieties of it each with its own specific function. In Hindu terminology the nearest equivalent to Qi is 'prana,' in Theosophy and Anthroposophy it is called the 'Ether' or 'Etheric Body.'

And an ancient definition, which proclaims the universal life principle:

> The root of the way of life (Dao or Tao) of birth and change is Qi; the myriad things of heaven and earth all obey this law. Thus Qi in periphery envelops heaven and earth, Qi in the interior activates them. The source wherefrom the sun, moon and stars derive their light, the thunder, rain, wind and cloud their being, the four seasons and the myriad things their birth, growth, gathering and storing: all this is brought about by Qi. Man's possession of life is completely dependent upon this Qi.
>
> Zhangshi leijing

According to acupuncturists Qi is present in the food we eat, the water we drink and the air we breathe. It is the continuous, unblocked flow of this life energy which carries the pulse of life, from the vibrating cells of our body, to every beat of our heart. Because the flow of Qi is now detectable with electronic instrumentation, some researchers explain this flow in terms of an electrical impulse (polarization) or wave of energy which travels along the nerves. The flow of Qi then, manifests in a polarity of negative and positive just as Chinese tradition has taught in the relative concepts of Yin and Yang. These complementary opposites are at the root of Chinese culture, pervading its art, literature, philosophy (especially Taoist) and medicine. When either becomes excessive, slowed or blocked, disease and even death may follow. Yin and Yang, then, must exist together in order for life to exist. They are but different aspects of the same basic life energy, Qi, which flows in our body most fully when there is openness and balance.

THE FIFTH ELEMENT: LIFE

Both Eastern and Western cultures recognize the four elements (which can be seen as cross-cultural, archetypal ideas): earth, air, fire, and water. The Chinese tradition recognizes these plus a fifth -- life -- out of which the other four evolve.

> The five elements, wood (life), fire, earth, metal (air), water, encompass all the phenomena of nature. It is a symbolism that applies itself equally to man.
>
> Su Wen

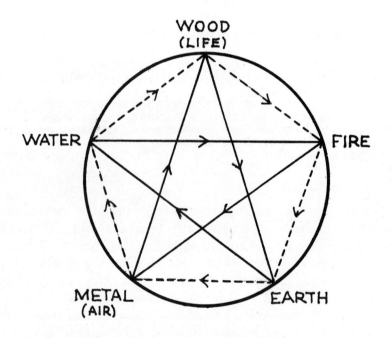

According to Chinese tradition, the five elements are related in two cycles of creation (around the pentagon) and destruction (around the star).

This fifth element has nearly been erased from our mind in the West. This is due to a materialistic bent in science toward explaining life as a consequence of "matter," rather than recognizing the life principle as the origin of "matter." This principle still survives in the concept of the universal field of vibration or ether whose existence Einstein was trying to establish at the time of his death. Most physicists have reasoned the idea of an ether or invisible carrier of light right out of their mind. Thanks to *Supersensonics*, however, the reality of the ether can be proven by anyone who will take the trouble to do so.

According to Chinese philosophy the life principle manifests first through the element of fire. So let's look at fire from the standpoint of modern physics and see if there is any life in it.

THE SUN IS ON FIRE

Present scientific terms for the four elements are solid (earth), liquid (water), gas (air), and plasma (fire). Plasma is a normally invisible gaseous, ionized field of charged or energized atoms and particle waves. According to some plasma physicists, plasma is present throughout and fills the entire universe, including our body. Plasma is susceptible to electromagnetic influences, can be shaped by magnetic fields, conducts electricity and can even produce light. The most spectacular example of plasma is our star, the sun, which is composed almost entirely of plasma.

SOME FAMOUS SUN GODS

No other heavenly body has been so widely worshipped as the sun. Entire religions have been founded on it, and for thousands of years the sun was the supreme god for many people. In this chart are listed seven of the most important sun gods in history, who worshipped them, and when.

Name	Where Worshipped	When Worshipped
Helios, (Apollo)	Greece	from about 3000 B.C. to A.D. 500
Ra	Egypt	from about 3000 B.C. to 1000 B.C.
Baal	Phoenicia	before 1400 B.C.
Mithras	Persia	from about 500 B.C. early history of Christianity
Amaterasu-Omikami	Japan	probably goes back 2000 years to the origin of Shintoism
Tonatiuh	Mexico	among the Aztecs, from their early history down to the 16th century
Vishnu	India	up to present among the Hindus

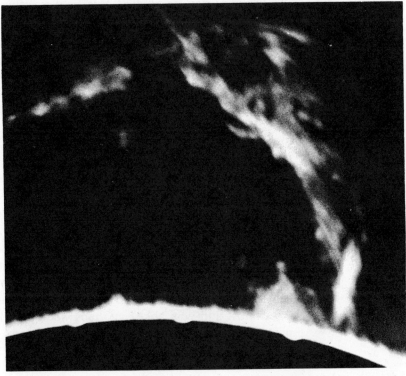

A solar flare arching along a magnetic loop, 200,000 miles above the surface of the sun.

OUR BODY IS ON FIRE

The bioluminescence revealed by electrophotography is leading some scientists to conjecture that this observable bioenergy is a phenomenon of bioplasma. That is, they believe that there is a "bioplasmic field" which interpenetrates all living things and acts as an energy pattern counterpart of our physical body. According to Soviet scientist Victor Inyushin:

> We submit that ultrafaint luminescence in biological systems represents plasmic processes. It's possible to somehow strengthen this ultrafaint luminescence in living things by the use of Kirlian photography. (18)

Inyushin has been investigating this phenomenon and feels that the bioenergy is not the mere radiation of our body; rather our body is the mirror-image of processes occurring in the bioplasmic field.* By using means other than electrophotography, he has avoided disturbing this field and has been able to observe its normal operation in us. His evidence convinces him that bioplasma in people is not a chaotic system, but rather is a specifically organized one which is different for different species. It includes ionized atoms and electrons, photons and other particles. He has found that this field is influenced by electromagnetic waves, and asserts that even solar flares and cosmic rays can cause changes in our bioplasma. It also radiates energy as well as absorbing it, just like the sun. Inyushin's research plus the phantom effect, Kim Bong Han's discoveries and Becker's success in regenerating limbs all point to an underlying energy matrix or cosmic blue print which is linked to dynamic processes occurring in all parts of our body. Several scientists now consider bioplasmic bioenergy to be very similar, if not identical, to the concept of prana in Yoga philosophy. For proof they point to the breath. Electrophotographic movies reveal that the brightness of the field corresponds with changes in breathing. (18) The bioluminescence or glow increases and decreases as we inhale and exhale. Breath, in fact, is the most significant physiological variable directly linked with changes in the observed radiation. This indicates that breathing is linked with ionization processes occurring in the bioplasma. In other words, when we breathe we not only inhale oxygen, we inhale life energy. Since prana, Qi or plasma (whatever we wish to call it), although normally invisible, is in every breath we take (according to scientist and yogi alike), perhaps we should look more closely at the process of breathing and the ionization processes that take place during oxygenation.

---

* See "Nuclear Evolution: Discovery of the Rainbow Body" by Christopher Hills, published University of the Trees Press, New Edition, 1977, for the body as cosmic hologram in a wavefield.

WE BREATHE FIRE

According to classical physics, every atom has one or more positively charged particles called protons in its nucleus and an equal number of negatively charged electrons in a spiralling series of concentric orbits around the nucleus. In the nonionized state, the two kinds of charges balance, and the atom is electrically neutral. But in many kinds of atoms, this balance can be upset and the atom then becomes charged or ionized. Some atoms are extremely stable. The most stable are the so-called inert gases: neon, argon, zenon, helium etc. Other atoms are not so stable (since they have slightly more or less than the stable number of orbiting electrons). They will readily gain or lose electrons until they reach that more stable state. When the loss or gain occurs, however, the positive charges in the nucleus are no longer balanced by an equal number of electrons and the atom becomes an ion with electric charge. A positive ion has lost one or more of its negative electrons. This charge is depicted in formulas as + (positive) or - (negative), the Yang and Yin of modern science, the basis of electricity and magnetism.

When atoms with these opposite tendencies appear together in water they transfer electrons readily. Water, therefore, is called a conductor. Since our body is composed of nearly 90% water, it is an all-pervading medium for the ionization processes occurring in the body fluids.

When we breathe, we inhale fiery energy in the form of air ions. The effects on our body of breathing negative and positive ions is proving to be considerable. As Beasley notes in Volume I, preliminary research shows definite correlations between the flow of ionic and electric currents in our body and the state of our physical and psychological health and, possibly, longevity. Negative ion generators are being used to eliminate illness, fatigue, grumpiness etc. due to the unnaturally high positive ion concentrations in urban and dusty environments. It seems that a balance between positive and negative ions in the air we breathe is essential to life.

OUR BLOOD IS ON FIRE

The word plasma is often associated with blood (that is, blood plasma). It is here, in our blood, that breath becomes life. Oxygen ions enter the aqueous solution of the blood stream after crossing the capillary membranes lining the alveoli or "air sacks" in the lungs. The blood acts as a carrier for the oxygen ions because the hemoglobin in the red blood cells contains iron which is a paramagnetic substance (can carry an electrical charge). This explains why blood (and possibly enzymes) is susceptible to electromagnetic fields. As we learned in

Volume I, "it is common knowledge that blood placed under a microscope with a magnetic field applied, reveals the blood cells all lined up in the direction of the field." In essence, our blood is on fire with the electrical charges carried by the hemoglobin.

OUR CELLS ARE ON FIRE

These electrically charged oxygen ions release their extra electrons across the double outer membrane of our cells. These electrons flow as cellular electricity in the liquid crystal conductors (cytochromes) of the cellular fluids. (6) The energy released through the oxidation process is a vital process of life, for it is essential for the manufacture of ATP, which carries energy to all parts of every cell. ATP (adenosine triphosphate) is a complex organic molecule which transports energy throughout our cells, as the bloodstream transports energy through our body. The production of ATP is controlled by the action of enzymes. Their production, in turn, is controlled by the hormones of the endocrine system.

Explained by Christopher Hills:

> The energy generating systems of all living things depend on two related actions in nature: oxidation plus synthesis of ATP. The action of oxidation releases electrons which act as agents for the transformation of energy. The action of ATP acts as a storage battery, feeding the cell with energy as it is needed. (6)

In the plant world these activities are regulated by the chloroplasts which are responsible for the photosynthesis of light, the source of all living matter on Earth and the origin of all biological energy. This photosynthetic process is mirrored in animal cells by the action of the mitochondria. There are hundreds of mitochondria in every cell of our physical body. They function like transistors, first attracting, next concentrating, and then amplifying these electrical flows. This energy is then released for electrochemical cellular processes including the production of DNA, the electrochemical messenger of life. The photosynthesis of DNA, like the photosynthesis of chlorophyll in plants, is dependent upon the energy exchange processes of oxidation.

The mitochondria act like nodal points (resonance bonds which interconnect us with the creative force-field of the universe) at the cellular level. They correspond to the acupuncture nodal points (acupoints) at the body level, and to the nodal points of the etheric force-field which surrounds and interpenetrates our body (see Volume I, the De La Warr research in biomagnetic resonance). Hills elaborates:

The molecular structure of the Mitochondrion is all summed up in the DNA nucleic acid with its message of replication and the ATP molecule which supplies the energy for that replication, including all the other processes of Life in the evolutionary thrust. Obviously, the replicating process is shaped by the guiding field of resonance radiation with oxygen atoms in the total environment. (6)

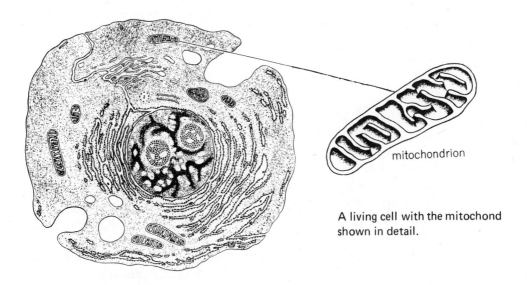

mitochondrion

A living cell with the mitochond shown in detail.

In other words, the mitochondria act like tiny semi-conductors and energy vortices which both absorb and emit energy necessary for cell life, including the creation of new cells or a whole new body (the structure of the DNA molecule is a spiralling vortex).* The degree to which the mitochondria vibrate in rhythm or harmonic resonance with the surrounding environmental field determines the ability of each cell to crystallize or trap the etheric life energy to which our body owes its physical existence, and by which we gain increased sensitivity to and awareness of life itself. The respiration of energy is a continuous function of every living cell. If a cell is deprived of this energy, even for an instant, it dies.

> Part of the process of "Nuclear Evolution" is to consciously learn how to operate the mitochondrion to draw the free electrons into our bodies from the earth's static field. Until now, only a few enlightened men have possessed this ability to attune their molecules and cells to the select wavelengths from the Lumen that would radiate their bodies in an excited state. A state of enlightenment/Nuclear Evolution, the words do not matter. Direct experience is the only true validator, and the only knowledge worth knowing. (13)

* The action of the mitochondria in drawing in energy is regulated by the hormones of the endo--crine system which is influenced by the chakra system.

HOW WE BREATHE FIRE

Ionization processes also occur in gases if strong electrical charges are present. Atoms in the air are ionized by electrical charges. The result can be a bolt of lightning. This explains the funny smell in the air in the vicinity of a lightning flash. Experimentally, a gas can be ionized by strong electrical charges in a partial vacuum tube, producing light. A good everyday example is the fluorescent light, a shining example of scientific technology which we will look at in Chapter Five. Little do we realize that a similar process occurs in our body with each breath we take.

Both the left and right nasal cavities are lined with a network of microscopic hair–like projections called cilia. Recent research indicates that the cilia are electrified and that one of their functions is to electrocute any bacterial or viral germs entering our body with the air. The cilia in the left nasal cavity are negatively (-) charged and those in the right are positively (+) charged, following the somatic polarities discovered by Dr. Albert R. Davis (noted in Volume I). It has recently been discovered that the nasal cavities in our forehead also act as ionization chambers for ionizing the hydrogen and oxygen atoms in the air we breathe. Although their concentration in the air is tiny (about one part per million) the effect of ionized hydrogen atoms in our body is considerable, for they undergo processes similar to those occurring in solar and stellar transformations.

> In the universe, hydrogen is apparently the most abundant element. Analysis of light emitted by stars indicates that most stars are predominantly hydrogen. For example, of the sun's mass, approximately 90 percent is hydrogen. (22)

The hydrogen atoms (and perhaps other atoms in the air we breathe) are ionized by the cilia releasing the one electron which orbits the hydrogen nucleus, leaving the single proton of the hydrogen nucleus with its positive charge. These freed electrons flow into the sphenoid bone, which encases the pituitary gland. According to acupuncture researcher Greg Brodsky:

> There is a large space in the sphenoid which acts as a chamber for the accumulations of the Ki which will nourish the pituitary gland. The sphenoid is therefore a protector of the pituitary gland as well as a "factory" of energy. (3)

In addition, the nerve which connects the eyes with the brain passes through the sphenoid bone. This may explain why acupuncture patients sometimes report an increased clarity of vision after treatment. Other bones forming the nasal cavities also appear to transmit Qi to other parts of the body. The ethmoid bone, for example, is attached to the membrane surrounding the brain through a projection on the bone called the *crista galli*. It apparently transmits energy to the entire brain. The maxillary bone at the side of the nose appears to collect Qi for the stomach as well as other parts of the body, for the stomach meridian begins there.

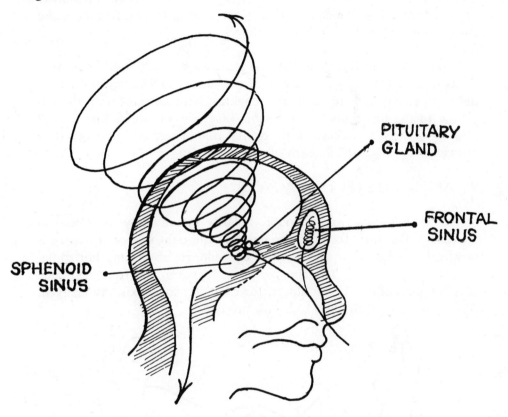

The hydrogen ions formed by the loss of electrons stripped away by the cilia flow through the bloodstream to all parts of the body. These ions are important in the formation of crystals from electrovalent substances, that is, substances which ionize. These substances are abundantly present as elements and liquid crystals in our cells. The ions unite in a lattice-like arrangement to make crystals. Hydrogen ions are essential to this process because they bind the different ions together. The structure and properties of proteins such as DNA and RNA depend on the existence of ionic hydrogen bonds. This is especially true in the four ventricles of our brain, where they enhance the protein synthesis in the cerebrospinal fluid which bathes the entire nervous system. These proteins act as miniature crystal sets which resonate with different frequencies of vibration depending on their structure. Thus the intake of hydrogen ions is essential to the proper formation of this liquid crystalline fluid. This directly influences our ability to ingest higher frequency cosmic energies because it affects our brain's ability to resonate with these vibrations. This in turn determines the level and quality of our conscious awareness of these higher energies constantly flowing through us. We can tune in to these energy flows, feel them and amplify them by watching our breathing.

If we watch our breathing the first thing we may notice is the rhythm of its flow in the rising and falling of our energy and strength. With the in-breath energy rises to the higher chakras. With the out-breath there is a flow to the lower ones; at the same time that we notice this vertical effect, we may also notice a horizontal one. There is an energy flow outward from the lungs on each side of our body.

WE ARE GIANT ELECTROMAGNETS

This flow results from a double polarity in our body. As noted in Volume I, Dr. A.R. Davis has found that the right side of the body is positive with respect to the negative left side. He also found horizontal polarities between the left and right, and between the front and back sides of the body. The shoulders, for example, are oppositely charged, front and back, much like a four-pole magnet.

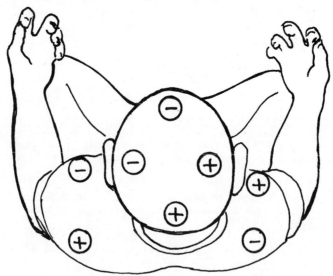

Our head is also electrically polarized left and right, front and back.

Whenever there is a difference in the potentials between two opposite poles there is a flow between them. In electrical terms this potential difference is called a voltage and the amount of flow between the poles can be measured as an electric current. Whenever there is a flow of electric current a surrounding magnetic field is generated at right angles to that current. There is, then, a double flow of current through our body. This is a higher form of energy than electricity however. Christopher Hills calls it the "psychic chakra electricity," which is more etheric than physical, although it has real physical effects.

You can test this out for yourself in a simple experiment. Simply stand erect with your back straight and arms stretched straight out level with the shoulders on either side of your body. Relax, breathe deeply and slowly clear your mind of wandering thoughts. Now focus your attention on your fingertips. Now slowly turn in a circle, your arms acting like the needle of a huge compass, your body. When you are aligned with the north-south magnetic field of the Earth, you should notice a sensation in the fingertips of the hand pointed north. This sensation depends on the person but may be a tingling feeling or a sensation of heat or coolness. If you continue to turn you will notice a similar but different sensation when your arms are aligned in an east-west direction. You are feeling the east-west flow of electrons which constantly flows around the planet at right angles to the magnetic polarity (see Volume III, *Supersensonics*, for a deeper explanation of these phenomena). With practice you will no longer need a compass to find direction; you can use the giant compass that nature has provided right here inside our body.

A SYMBOL OF LIFE'S FIRE

The caduceus of Mercury, sometimes called the staff of Hermes, is an ancient mystical symbol for life and health. Its origin has been traced back to the physicians of ancient Egypt and may have been brought to Egypt by descendants of the legendary civilization of Atlantis. A nearly identical symbol appears in Yoga philosophy where it is called *meru danda* in Sanskrit.

The caduceus is still used today as the symbol for medicine in pharmacies and hospitals everywhere. Unfortunately however, very few people connected with the medical profession can explain what it means or how it can be of use to us because medical schools long ago stopped teaching the wisdom symbolically written in the image. It has been ignored and forgotten. This may be due to the materialistic bent in medicine which assumes that disease is best cured with physical means such as synthetic chemicals and serums, complicated machines and costly and often traumatic surgical operations. Medicine is big business now and treatment is increasingly related to the size of one's purse. The

In the Bible Moses prayed to God on behalf of his people who were dying from snakebite. He was told to "Make an image of a fiery serpent and set it on a pole; and everyone who is bitten, when he looks upon it, shall live." So Moses made a serpent of brass and lifted it up on a pole; and when anyone was bitten they looked upon the serpent of brass, and lived. (Numbers 21: 4-9)

The caduceus of Mercury (Roman) or staff of Hermes (Greek).

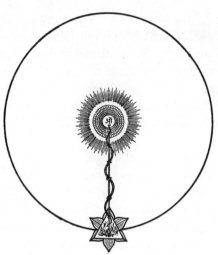

The meru danda, giving rise to the Spiritual Sun.

Aesculapius, the Greek and Roman god of healing

Aesculapius' Staff modified for use as a symbol by the U.S. Department of Health, Education, and Welfare

prescription forms and apothecary jars which carry the caduceus on them make a mockery of what this symbol really means as an aid to understanding the healing process.

Biblical references to the trees in the Garden of Eden are analogies for the meaning hidden in the caduceus.

> And out of the ground made the LORD God to grow every tree that is pleasant to the sight, and good for food; the tree of life also in the midst of the garden, and the tree of knowledge of good and evil.
> -- Genesis 2:9

This meaning becomes clear when we see the Garden of Eden as our consciousness and the trees of life and knowledge as the energy flows through the nervous system of our body, which can be seen as a huge tree that is upside down. Instead of in the ground the roots of the nervous system lie in the sky, the brain; the spinal cord is the trunk and the branches are the nerve fibers which stream out to the extremities. The nervous system is the physical vehicle for the manifestation of mental processes, for the crystallization of mental energy into matter. As it is with all energy, mental energy is polarized into positive and negative currents which mold and activate our body. This manifests in a four-fold system of interconnected forces following the electromagnetic polarities outlined earlier. At the positive source or center of consciousness at the crown chakra the energy is brilliant, superconscious and of high frequency and intensity. As this radiant energy flows outward and downward through our nervous system, it splits into two polarized streams which meet at the base of the spine. At the end opposite the source, the negative pole at the base chakra, it becomes crystallized into matter. This descent of mind into matter in search of sensation has been called the process of involution. The Biblical symbol for the mind is the serpent, the great deceiver. This is the serpent that came down the tree of knowledge to tempt Eve (the begetting, generative, shaper of forms, Shakti, divine mother principle present in all beings whether male or female).

This entire process is symbolized in the caduceus. The knob or bulb at the top of the staff represents the optic thalamus and pineal gland which act as the center of consciousness and psychic light in the brain. The double wings symbolize the two brain hemispheres, each split into primary and secondary parts. The two large wings are the cerebrum with its billions of nerve endings, the ruler of the cerebrospinal system, the tree of life in Genesis. This is the thinking center, the dynamic functioning part of the brain where all electrical impulses generated by mental processes are directed and correlated. The two smaller wings represent the cerebellum which regulates the automatic functions of our body through the sympathetic system. The central shaft is the spinal cord in the spinal column. The two serpents represent the mind principle in its dual aspects passing downward through the sympathetic system, which consists of a double chain of ganglion originating in the brain and ending in front of the coccyx. Each chain is symbolized by a serpent and where they cross over denotes the various nerve plexuses which link the sympathetic system with the cerebrospinal system. The subtle nerve currents change polarity at these crossover points, flowing constantly in and out of each other and producing alternating currents of psychic electricity.

The flow of breath through the nostrils into the lungs is mirrored by the flow of psychic electricity through the subtle nerve pathways which are called nadis in Yoga. Both nadis begin at the root of the nose, the *ida* on the left and the *pingala* on the right. The ida and pingala are said to diverge into many small branches and subbranches which unite all the cells in the body with the surrounding ether via the etheric field. They represent the basic pattern of psychic electrical flow into our nervous system and acupuncture meridians. Ancient Yoga texts speak of the nadis as wound like two spiralling, intertwined snakes along the spinal column, just as modern science has confirmed in analyzing the nervous system.

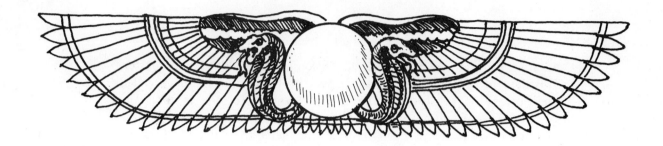

The Sun Disc, an ancient Egyptian symbol for the generation of life's fire (the sun) through the balancing of the positive and negative forces (the two cobras) giving rise to eternal life (the spread wings). This symbol is very similar to and may have been the origin of the ancient Greek staff of Hermes and later Roman caduceus of Mercury.

As it enters the etheric forcefield of our body, prana is split into polarized energies: positive, or *Shiva* energy to the right, and negative, or *Shakti* energy to the left. Shiva represents solar, masculine, expansive energy called Yang in Chinese philosophy. Shakti is lunar, feminine, contractive, Yin energy. The way we breathe prana or psychic electricity is just as important as the way we breathe air. If we are seldom fully aware of breathing, how often are we even dimly aware that we are breathing energy (prana)? Perhaps we can recall an experience at the beach, atop a hill, at the base of a waterfall or on a still, frosty morn when the air was so supercharged with energy that we felt renewed and filled with life with each fiery breath!

WE ARE DYNAMOS

Our body is deeply influenced by the flows of psychic electricity (prana) which pass through us. These flows follow the same basic principles as electric currents and can be explained in terms of electrical conductivity. (21) The process of generating electricity can be represented as a sine wave. By a phase shift of 180 degrees, the impulse returns from the end. This is called a standing wave.

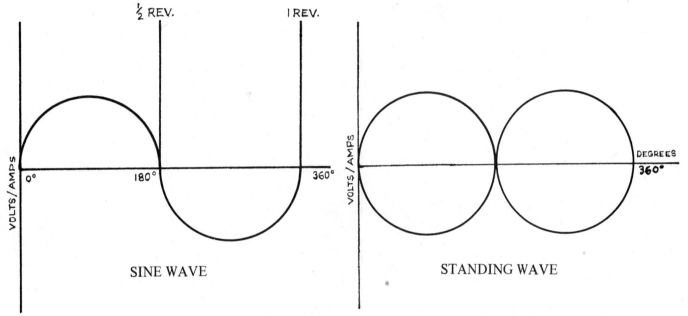

SINE WAVE   STANDING WAVE

Basically, this is how the psychic nerve pathways act in our etheric body. They form a double spiral standing wave with the flows of positive (Shiva) and negative (Shakti) psychic electricity. These currents seem to cross at points between the chakras. The chakra energy centers arise as a result of these positive and negative flows of psychic electricity. Each chakra acts like the rotor of an electric motor which revolves when negative and positive currents flow around it. (21) When the polarity of the current reverses, the rotor changes the direction of its rotation.

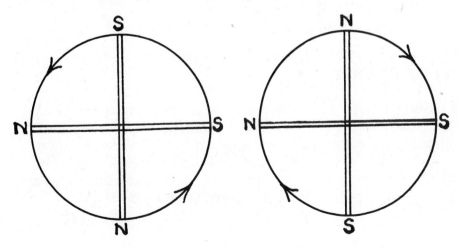

An identical process occurs at each chakra. The chakras act like turbines for absorbing prana and free electrons from the east-west flow in the atmosphere.

Though it is not generally recognized, all the electricity we get from our electric power generating plants is captured from this east-west flow of electrons. The energy applied to turbines (falling water, steam, gas) turns them inside a magnetic field. This creates a magnetic vortex which, offering less resistance than the atmosphere, sucks in the free electrons converting them into a flow of electricity. The chakra energy centers, then, can be likened to psychic turbines which absorb and radiate energy of different wavelengths or frequency bands depending on their speed, direction and angle of revolution. Our chakra energy system functions like a superconducting electrodynamo.

THE FIRE THIS TIME

In addition to devices which trap and harness the energy from the sun in the form of solar collectors, there is another solar technology which duplicates the processes occurring in the sun itself. This technology of plasma physics is called magnetohydrodynamics. It deals primarily with recreating the power of the sun in the laboratory. A special device has been invented which is called the magnetohydrodynamic shock-tube generator. Basically, this consists of an ionizing gas (plasma) encased in a tube or magnetic bottle surrounded by strong electromagnets and connected with electrical circuitry. When an electrical shock is passed through the tube it excites the atoms of the plasma to a higher energy state. This produces energy in the form of electrons, photons and other particles. The more the plasma is shocked the stronger the magnetic field becomes, for it acts like a vessel to contain the reaction. Preliminary research has shown that this process is very efficient as an energy producer.

We now have proof that the processes of the sun can be reproduced mechanically. If we can do this with a machine, why can't we create a small sun in the laboratory of our own body? After all, the major constituents are here. We have electric and ionic currents flowing throughout our body and inside every cell. These currents generate interconnecting magnetic fields. As we learned in Volume I, our spinal

column is bounded at both ends by a strong positive charge, the spine itself having a negative charge. It is filled with electricity flowing in the nervous system. Our body, some scientists say, is interpenetrated by a bioplasmic field which causes the glow seen in electrophotography. Hydrogen atoms, ions and isotopes of hydrogen are present in our body, though in smaller concentrations than the sun. These processes all occur in an aqueous solution, therefore the heating effects of solar processes may be muted by the watery nature of our physical body. Experiences of intensely uplifting mental, emotional and sensory excitement sometimes bring a feeling of a current of energy running up and down the spine, even suffusing our entire body with blissful energy. There seems to be some connection, therefore, between an open, loving heart and the experience of pure joy radiating from our innermost, nuclear center of being. Are we potentially a star, radiating our warming, healing radiance in all directions?

KUNDALINI -- THE SUN WITHIN

The Caduceus of Mercury is sometimes depicted with a snake coiled three and one-half times at the base of the central shaft (sushumna in Yoga) representing the spinal column. This snake symbolizes the "serpent fire" of kundalini energy, or the Akasha, our latent potential for purified and expanded states of Pure Consciousness which lies waiting inside each of us. Once released from the blocks which impede it, it doubles its length as it uncoils up the spine, finally connecting our individual consciousness with the guiding logos, or divine plan of the universal intelligence, through the crown chakra.

When we learn to consciously "tune in" to our breathing (exercises are provided at the end of this chapter) we can begin to actually *feel* the flows of these subtle, dynamic energies. We can experience the psychic electricity flowing through the chakras, our nervous system and even the acupuncture meridians. This practice eventually brings with it the natural ability to balance the horizontal and vertical polarities by finding and removing the blocks which stifle their flow. Finally, through continued practice we can balance them so that they cancel each other out. This creates a dynamic stillness at the center of our being (a sort of psychic homeostasis) which is undisturbed by "outside" conditions. According to Christopher Hills:

> Kundalini is the flow of consciousness unobstructed by limitations in the mind-stuff controlling the chakra system. This flow of psychic electricity runs through the human organism and guides its evolution of the Nuclear Self, quite independent of the actual cells and organs which absorb the psycho-physiological energy from the cosmic light radiation which is always present in biological activity. (7)

Kundalini (or consciousness) flows naturally as a result of transcending the mind and thoughts. To attain this, the incessant flow of ideas and mental images must stop, mental tapes must be erased and all mental trips have to be cancelled. When we can realize that I the thinker am independent of thought we can stop thinking and remain conscious at the I am level alone. No longer asking the question – Who am I? – we contact *That I am,* the One who is All. This opens the eye of wisdom, the prime imagination of the universal intelligence, through the crown chakra. This brings experience of more purified and expanded states of consciousness – supreme being – through the stilling of the restless distractions of the senses and thoughts. Although kundalini normally flows erratically and not past the first or second chakra in most of us, it can be raised to the crown chakra. Fully opening this center rolls the wheel that stops the worlds of time and space, fulfilling the Biblical injunction:

WHEN THINE EYE BE SINGLE, THY WHOLE BODY WILL BE FILLED WITH LIGHT.

The course *Into Meditation Now* is designed to activate the kundalini fire of Pure Consciousness. The goal is direct enlightenment, based upon original insights by a master of consciousness into the teachings of Christ and other great emanations from the source of us all. This course is practiced daily by the student-faculty of the University of the Trees who are committed to eliminating the illusory barriers which separate us from one another, and our True Self. This meditation course is available by correspondence to all those prepared to face the real teacher or guru – life itself.

THE KUNDALINI ROOMPH COIL

The word roomph is a combination of the Sanskrit word Rum which means spirit and the American word oomph. This is the kundalini energy which rises up the spine through the chakras and gives us oomph. Many of us are blocked from using this power because of lack of control through inadequate spiritual discipline. The Kundalini Roomph Coil is designed to help us gradually release this transformative energy. This is a safe way to raise the kundalini flow without damage to the spinal nerves and chakra energy centers through lack of preparation.

> Gradually you raise your Roomph until you begin to feel very sweet inside and a bubbling sensation begins to awaken a new quiet joy within your heart. This sweet feeling has little to do with what is happening in your external life. Things can be falling down all around you in chaos but you remain peaceful, collected, and quietly excited by this continuous flowing inner radiance. (8)

SEVEN STAGES OF GROWTH

The most important aspect of prana is the flowing movement of the breath. Since this movement is the manifestation of life energy, the energy can be controlled by controlling the breathing. Breathing directly affects the flow of energy in the sushumna. There is a flow when the breath alternates between the left and right nostrils, when both are flowing equally and when neither is flowing at all. The first flow derives from a continual balancing of unequal energy. The second arises from a balance of equal, but opposite, energies. The third arises from the first two, plus their balanced cessation during retention of the breath. This energy flow has been described as the "sap" in the tree of knowledge (the nervous system). However, if we have many blocks and filters that limit our awareness, then they will be magnified by this energy to the point where there may be a tremendous force running out of control. This is the danger of undisciplined release of kundalini which sometimes comes from unwise practices. The basic steps of control and discipline must first be understood and undertaken if our desire is real self-mastery. Otherwise, it is like putting a thousand volts through a five volt circuit — we will burn out our circuits.

As we learned earlier, Yoga is a mystical science based on seven stages of growth which reflect the control of life energy at the seven chakra energy centers. We begin with the firm determination to investigate our mental nature — how the mind works. There are five resolutions *(yamas)* the practice of which enables us to gradually get our thoughts under control. They include:

The will neither to kill nor to injure.

The will always to speak the truth and to act truthfully.

The will to always remain honest and not take what is not given.

The will to direct all attention and energy toward realizing Ultimate Reality.

The will to seek or take nothing for personal gain.

Without a firm foundation from these basic resolutions little progress can be expected in the pursuit of Truth. In fact much danger can be expected, for failure to follow them will only expand our problems later on.

The first step is then combined with *niyama*, which is the firm direction of mental and physical activity toward Truth by practicing:

Cleanliness of body and mind.

Contentment.

Critical examination of the senses.

Study of physics, metaphysics, psychology and books of self-knowledge.

Realization of the oneness of all existence.

The third technique, *asana*, brings control of the body and elimination of blockages, allowing freer flow and expression of life energy. The physical discipline of hatha yoga exercises automatically develops control of the intellect over the body.

The fourth technique is *pranayama*, the aim of which is to regulate the flow of inner breath by learning to breathe inwardly (exercises for which come later). The fifth stage is *pratyahara*, where energy is mentally withdrawn from the lower chakras and concentrated in the highest one. This technique is only effective when the previous four have been mastered. Next comes *dharana*, where the mind is fixed one-pointedly on a particular object or idea. This brings an experience of oneness, when the object of concentration becomes identical with self in self-concentration. Finally, there is *dhyana*, the total expansive experience of spiritual union which brings self-mastery and direct perception of Truth.

NUCLEAR EVOLUTION

It has long been a basic assumption of classical physics that we must be impartial, "detached" observers of "external objective reality" and that valid experiments should be repeatable by any competent observer. Yet, if we could produce a photograph of the entire planet Earth showing us where everything is, it would only be true for an observer when the picture was taken. A photograph taken a second later would reveal an entirely different world. When we lie down to sleep at night, do we realize that we do not awake in the same world, or even in the same body? All around us everything is in constant flux. Even the atoms and molecules composing "inanimate objects" are alive with vibration, undergoing incessant subatomic and nuclear transformation. Change is the only constant. Do we resist change, or do we learn to understand change, welcome it and even rejoice in it?

Each of us is a pivot in the entire cosmic evolutionary process. Everything we think, say and do affects everything else. We hold direct responsibility for the evolution of all life on Earth. We can only influence or change life, though, by first understanding and changing ourself. We can expect little success by telling others to change. If we truly wish to make this a better world we must begin where we can be sure of success -- with ourself.

Change is the essence of nuclear evolution. We can not change anything we are not aware of, so a change in our consciousness is the first step. To achieve this we begin at the Nux -- the very center of our being, and all being.

> Our spirit which is the highest expression of ourself, sits inside our body guiding our evolution, in much the same way as a seed sits inside its shell and contains the blueprints for the unfoldment of the total flower. We refer to this seed as our Nux, a Latin word meaning nut or kernel. It is our true self sitting inside this Nux that we wish to evolve. (13)

Nuclear Evolution is a model for life, and a way by which we can evolve to Cosmic Consciousness, by understanding our origin -- that which created us. We will not be able to experience the Nux, or nuclear center of our real being, until we see through our self-sense or egocentric notions that we are a separate self (which manifests as onlooking, observer consciousness). Everything is only external to an observer who separates self from what is observed. Everything is internal to a participator who joins the center of self with the center of what is observed. Only then will we be able to cut through the appearance of things to the underlying essence of pure being. This is center consciousness at work. To experience reality from the center of a system is to be in resonance or rhythmic harmonic attunement with everything in that system. Center consciousness ends the separation between the experiencer and what is experienced, thus cutting through the different velocities, drives and time-worlds to the still, timeless essence of pure being.

Through an understanding of nuclear evolution we can hasten the evolution of consciousness in ourself. By consciously absorbing more and more energy (light) into the center of our etheric field (aura), we raise all the cells in our body to a higher energy state -- a higher dimension of being. This increases our bioenergetic sensitivity to the full spectrum of that universal field of vibration, from zero to infinity, which interpenetrates and creates our entire being. This brings about the purification and expansion of our consciousness which is necessary for the fruitful practice of the ultimate science of center consciousness -- Supersensonics. We will look more deeply into Hills' theory of consciousness -- nuclear evolution -- in the next section.

EXPERIMENT SECTION FOR CHAPTER THREE

BALANCING THE BREATH OF LIFE

Our body is like a miniature universe, animated by the breath. The aim of hatha yoga asanas is to equalize the breath. The "ha" or solar, positive, electrical nervous energy flows in the pingala and the "tha" or lunar, negative, magnetic, fluidic energy flows in the ida. When both are balanced, the kundalini flows in the sushumna.

The normal flow of breath follows a succession of cycles found in swara yoga, the experiential science of consciously patterned alternate nostril breathing. In a person of normal health there is an average of fifteen breaths per minute, or 21,000 breaths each day. This follows a natural rhythmic cycle of alternate nostril breathing with a short balance in between. The whole cycle of breathing each day follows approximate two-hour cycles of left, then right nostril breathing. These twelve natural divisions correspond with the twelve signs of the zodiac. As the sun passes into the sign Aries, more prana flows into the energy pathways of the brain. During the day, the concentrated flow descends along the spine through the areas of the respective signs. Before midday the effect is primarily solar (nervous). After the balance at midday the effect is primarily lunar (fluidic). This occurs because the first six periods of increasing light are controlled by the sun and those afterward by the moon. Midday and midnight are the two times of greatest balance between right and left, positive and negative and solar and lunar, which bring a stronger flow in the sushumna. Although these flows occur naturally the effects can be considerably enhanced through conscious control of breathing.

The conscious control of breathing comes through pranayama. By controlling the flow of breath we control the manifestation (yama) of life-energy (prana) in our body. In normal breathing at the rate of about fifteen breaths per minute we inhale and exhale only about one pint of air with each breath. If we inhale more deeply we can take in three additional pints of air. If we exhale more forcefully with the diaphragm we make room for another three pints. Through gradual breathing practice the capacity of the lungs can reach nearly ten pints. This means, then, that we can potentially increase our intake of oxygen and air ions by ten times. The deep breathing of pranayama supercharges the respiratory, nervous, venous, and meridian circulatory systems with life energy. This increases each cell's ability to ingest life energy and excrete waste products, which in turn brings prolonged and more energized life, as well as heightened sensitivity of our entire body.

There are four stages in the breathing cycle: inhalation, retention, exhalation, retention. In pranayama, these four processes follow a rhythmic pattern in each exercise. The upper chest, thoracic area, and diaphragm region are all used in pranayama. In most of us, it is the diaphragm area that we need to work on in order to increase our intake.

THE ABDOMINAL LIFT

The abdominal lift (Uddiyana Banda in Sanskrit) is considered one of the most essential hatha yoga exercises for both our physical and our psychical development. It strengthens the abdominal muscles and helps build a "flat tummy." It brings relief from gas, constipation, indigestion, and liver trouble. Yogis practice this exercise to develop spiritual energy, for it tones up those nerves which have their roots in the solar plexus region. (CAUTION: The abdominal lift should not be attempted by anyone suffering from a weak heart, serious abdominal or intestinal problems, or pregnant or menstruating women.)

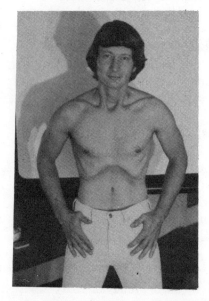

TECHNIQUE: Stand with your feet about one foot apart, inhale deeply and exhale all the air with force. Then, without inhaling again, draw in the abdominal muscles with a strong upward pull until a hollow forms under the ribs. Place your hands on your thighs, bend the knees a little, and slightly tip your trunk forward without lowering it. The diaphragm then rises easily. Keep your hands pressed firmly against the thighs when leaning on them. Stay in this position as long as you can without breathing. Relax, inhale and stand up straight. When your abdominal muscles become stronger through daily practice, try alternately relaxing and clenching them instead of just holding them in. This will strengthen them even more.

BREATHING EXERCISES

Our body is a temple for the living light of consciousness. Therefore we need to observe cleanliness of both body and mind. Pranayama, like hatha yoga, should be practiced with an empty stomach, an empty bladder and empty bowels. This eliminates discomfort, possible injury and internal obstructions to the flow of life energy. The best times of practice are early morning and early evening, though it only takes a few minutes and can be done almost anytime, anywhere (it is especially effective in removing headaches). The techniques of pranayama join the mental faculty of awareness with flows of life energy in our body. By merely being aware of the breath (not actively thinking about it), it miraculously changes by itself. It slows down, becomes longer and deeper and floods our subtle energy circuits (nadis) with the fiery essence of life. By removing mental and emotional blocks which inhibit its flow, we can not only feel this powerful life energy but also *heal* ourself with it. Pranayama, then (practiced regularly) has deep healing effects, especially to counteract the pollution that enters us when we breathe shallowly and unconsciously. By learning to become more fully aware of the breath, we become more conscious of life. The following exercises are adapted from *Wholistic Health and Living Yoga* by Malcolm Strutt (available from the publishers).*

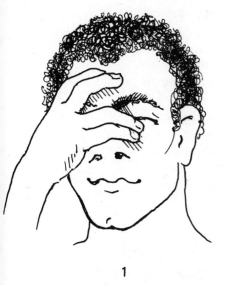

1

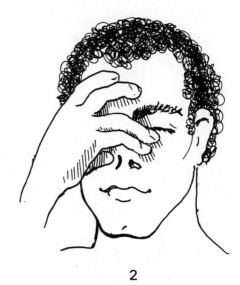

2

STEP ONE: Begin in a comfortable sitting position with the back erect (but not rigid) and spine straight. Place the right hand and fingers as illustrated (figure 1).

STEP TWO: Press the right nostril closed with the thumb and inhale slowly and quietly through the left nostril to a count of seven beats (figure 2).

STEP THREE: Close both nostrils and retain the air for seven beats (figure 3).

STEP FOUR: Release the right nostril (still holding the left closed) and exhale slowly and quietly through the right nostril to a count of seven beats (figure 4).

STEP FIVE: Keep the left nostril closed and without pause, inhale slowly and quietly through the right nostril to a count of seven beats.

STEP SIX: Close both nostrils and retain for seven beats.

STEP SEVEN: Release the left nostril (keeping the right closed) and exhale slowly and quietly through the left nostril to a count of seven beats. Repeat this complete cycle seven times. (CAUTION: Do not retain the breath if you have high blood pressure.)

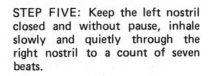

3

4

This exercise has a profound effect on the mind-body relationship. An angry, emotionally aroused or frightened person will breathe rapidly. Obviously then, watching the breath induces the opposite effects of relaxation and a tranquilizing balancing of mind and body. This control also brings the physiological effect of a more harmonized metabolism.

* This is an in-depth, practical course of instruction for advanced students of hatha yoga. It combines yogic knowledge of life energy flows with conscious awareness for the understanding and integration of the seven levels of consciousness of nuclear evolution.

In order to add natural rhythm to pranayama, we next follow the prime mover of the life energy in our body -- our heart. We can tune the rhythm of alternate nostril breathing to the beat of our heart by: one, feeling the pulse in the wrist, two, listening with a stethoscope, or three, tuning in to the throb of the pulse anywhere in our body, through concentrated awareness. This last method is especially good for expanding consciousness if we concentrate all our awareness on the area above, between and behind the eyes (the Ajna or brow center) as we breathe. With each inhalation concentrate on feeling energy rising to that magnetic and intuitive center of mind. During retention feel it expand, like a sun radiating in all directions, with each beat of the heart. The nadam or supersensonic sound current may arise spontaneously as a ringing or buzzing in the brain or ears. If it does, concentrate on that while continuing to practice pranayama. The experience is well worth the effort, for it weds consciousness with energy and expands both as they become one.

RUMF ROOMPH YOGA

Rumf Roomph Yoga is a science of vibration and transmission of life energy based on a series of oral teachings recorded on cassette tapes available from the publishers. The essence of Rumf Roomph Yoga is contained in the vibration of the mantram:

RUMF ROOMPH DRIVENAM SWA

RUMF means spirit of life, as in the Sanskrit word Rum meaning "spirit."
ROOMPH means the pulsating vital force of Prana bubbling up from the base of the spine and spilling over into the heart, as in the English word "oomph," combined with the Tantric word ROOM. "Tremendous enthusiasm" could be another interpretation.
DRIVENAM means the wealth of the universe or total security of the Cosmic Credit Card, but here more to be interpreted as a state of being beyond and out of the mind, into the real self and penetrating the entire being with a radiant beneficence.
SWA means sweetness or bliss which, like ambrosia, melts the Ego, as a baby does, until the self becomes sweet inside its own self, producing saintliness.

Rumf Roomph Yoga is concerned with making every cell in the body bubble with life energy and using that energy for practical purposes, such as growing new brain cells, raising the concentration of positive hydrogen ions in the cerebrospinal fluid for expanded perception and improved protein synthesis and stilling the restless distractions of the mind and senses. In short, it is a guide to the experience of ultimate reality through practical techniques that really work.

CENTERING

In this exercise, we add a visual concentration device, the powerful mandalic center symbol of nuclear evolution. It is specially designed to unite the negative/positive, Shakti/Shiva, Yin/Yang energy flows by tuning in to the essence behind the mind stuff which generates these flows and balances them by turning the mind back on itself, stilling it and ultimately transcending it.

STEP ONE: Sit in a comfortable position with spine erect and back straight about three to four feet from the center symbol at the end of this chapter. The symbol should be at eye level. Focus all your attention onepointedly at its center. Breathe deeply and slowly to aid relaxation and concentration. Gradually you will see spirals spinning both left and right which will begin to produce a special silver sparkle of white light circling around the central core of the symbol. Visualize the symbol turning into a tunnel which comes out around you until you feel you are inside it walking toward that center of blazing white light.

STEP TWO: Perform step one, but this time sit in a darkened room which is illuminated by only a blacklight long wavelength, ultraviolet fluorescent light which is placed near the center symbol to light it up. Shield the ends of the bulb with thin lead strips to absorb the radiation which is emitted by all fluorescent lights. This will help to correct the deficit of this important, health giving vibration of light which is presently lacking in our artificial environments. It will also raise the vibration of the center symbol considerably.

STEP THREE: Sit in a comfortable position and commence alternate nostril breathing with your eyes focused on the center symbol. During each period of retention close your eyes and concentrate on the after image of the symbol in your mind's eye. At the beginning it will gradually fade away outwards and upwards. Through continued practice you will be able to hold the image longer and longer. It will gradually turn into a tunnel of light and you may begin to feel energy streaming in through the crown chakra at the top of your head which the center symbol not only represents but also activates. This exercise gradually brings about the experience of a bright light in the head, a psycholuminescence that is like a sun, radiating in all directions. This exercise is partially extracted from session number five of the Rumf Roomph Yoga series. Session number five takes you on a voyage much farther down this tunnel of self-discovery than the first steps given here.

SHUTTING OFF THE SENSES

Finally to put all this energy and mental concentration to practical use, we add the yoga posture called yoni mudra. Yoni means womb or source - the Absolute which is the heart, the seed-center for all existence. Mudra means a physical practice which has an effect on the mind. Basically, this exercise effectively and simply cuts off the senses and leads to a semiautomatic state of pratyahara (trance), especially when done during the long retentions in the alternate nostril breathing exercises.

Sit in a comfortable position, again with the spine erect and back straight (but not rigid). Close the ears by inserting the thumbs. Close the eyes with the forefingers. Place the middle fingers on either side of the nose, at the bridge. Close the upper and lower lips with the ring and little fingers, respectively. Commence alternate nostril breathing, using the middle fingers to open and close the nostrils.

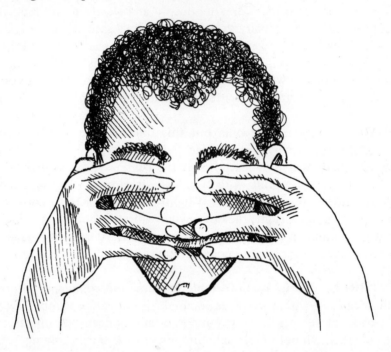

This exercise shuts off the mind from extraneous sensory stimulation and "outside" interference. It results in an almost automatic state of pratyahara (trance induction through sense withdrawal) by both balancing the flows of psychic electricity in the subtle nerve channels (nadis) and raising that energy up the spine to the higher energy centers. This upward flow of energy which comes from stopping the breath, halting the flows of nervous energies from the physical senses and mental thoughts and balancing the chakras at resonant nodal points of stillness, is kundalini -- the light of consciousness. Pratyahara, when combined with one-pointed concentration (dharana) and the continued suggestion of union in meditation (to the exclusion of all other thought), brings us ever closer to that divine spiritual union of inner with outer, real with ideal, micro with macro, individual consciousness with Cosmic Consciousness, self with Self.

Pranayama, the science of breath, is also the science of life. With pranayama we begin a new creation linking our breath with that Great Breath which moves all things. There is a nuclear seed-center in each of us that lies in the black void beyond the barrier of light that binds us in the seven skins of individual consciousness. By generating that light and radiating it in all directions like a sun, we vibrate our consciousness at a thousand times the speed of light. Only then is the light barrier breached and the latent potential of transcendental vision manifest as the clear light of PURE CONSCIOUSNESS!

REFERENCES FOR SECTION ONE

BOOKS

1 Argüelles, José and Miriam. MANDALA, Berkeley: Shambala, 1972.

2 Blair, Lawrence. RHYTHMS OF VISION, New York: Schocken Books, 1976.

3 Brodsky, Greg. FROM EDEN TO AQUARIUS, New York: Bantam Books, 1974.

4 Brown, J.H.U., and Baker, S.B. BASIC ENDOCRINOLOGY, Philadelphia: F.A. Davis, 1966.

5 Capra, Fritjof. THE TAO OF PHYSICS, Berkeley: Shambhala, 1975.

6 Hills, Christopher. NUCLEAR EVOLUTION: DISCOVERY OF THE RAINBOW BODY, (second edition). Boulder Creek, CA, University of the Trees Press, 1977.

7 Hills, Christopher. SUPERSENSONICS: The Spiritual Physics of all Vibrations from Zero to Infinity. Boulder Creek, CA: University of the Trees Press, 1975.

8 Hills, Christopher. SUPERSENSONIC INSTRUMENTS OF KNOWING, Boulder Creek, CA: University of the Trees Press, 1975.

9 Krippner, Stanley and Rubin, Daniel (eds.) THE KIRLIAN AURA: PHOTOGRAPHING THE GALAXIES OF LIFE, Garden City, NY: Anchor Books, 1974.

10 Lane, Earle. ELECTROPHOTOGRAPHY, San Francisco: And/Or Press, 1975.

11 Luscher, Max. THE LUSCHER COLOR TEST, Translated and edited by Ian A. Scott. New York: Pocket Books, 1971.

12 Massy, Robert. ALIVE TO THE UNIVERSE: HANDBOOK FOR SUPERSENSITIVE LIVING, Boulder Creek, CA: University of the Trees Press, 1976.

13 Massy, Robert. HILLS' THEORY OF CONSCIOUSNESS: An Interpretation of Nuclear Evolution, Boulder Creek, CA: University of the Trees Press, 1976.

14 Mishlove, Jeffry. ROOTS OF CONSCIOUSNESS, New York: Random House, 1975.

15 Moss, Thelma. THE PROBABILITY OF THE IMPOSSIBLE, Los Angeles: J.P. Tarcher, 1974.

16 Offner, Franklin. ELECTRONICS FOR BIOLOGISTS, New York: McGraw-Hill, 1967.

17 Ott, John N. HEALTH AND LIGHT, Old Greenwich, CN: Devin-Adair, 1973.

18 Ostrander, Sheila and Schroeder, Lynn. HANDBOOK OF PSYCHIC DISCOVERIES, New York: Berkeley Publishing, 1974.

19 Ostrander, Sheila and Schroeder, Lynn. PSYCHIC DISCOVERIES BEHIND THE IRON CURTAIN, New York: Bantam Books, 1971.

20 Presman, Alexander S. ELECTROMAGNETIC FIELDS AND LIFE, New York: Plenum Press, 1970.

21 Rendel, Peter. INTRODUCTION TO THE CHAKRAS, New York: Samuel Weiser, 1974.

22 Sienko, Michell J., and Plane, Robert A. CHEMISTRY, New York: McGraw-Hill, 1961.

23 White, John (ed.) THE HIGHEST STATE OF CONSCIOUSNESS, Garden City, NY: Doubleday, 1972.

ARTICLES

24 Hameroff, Stuart Roy, "Ch'i: A Neural Hologram? Microtubules, Bioholography, and Acupuncture." American Journal of Chinese Medicine, Vol. 2, No. 2 pp. 163-170, 1974.

25 South, John. "New Technique Can Detect Pin-sized Cancers Years Earlier Than Present Methods." National Observer, Nov. 2, 1976.

26 Staff, New York Times. "Fluorescent Light's Effect on Cells." San Francisco Chronicle, April 28, 1977.

SECTION TWO

MIND AND MATTER

Seeing. We might say that the whole of life lies in that verb -- if not in end, at least in essence . . . union can only increase through an increase in consciousness, that is to say in vision . . . After all, do we not judge the perfection of an animal or the supremacy of a thinking being, by the penetration and synthetic power of their gaze? To see more and better is not a matter of whim or curiosity or self-indulgence. To see or to perish is the very condition laid upon everything that makes up the universe, by reason of the mysterious gift of existence. And this, in superior measure, is man's condition.

Pierre Teilhard de Chardin, "The Phenomenon of Man"

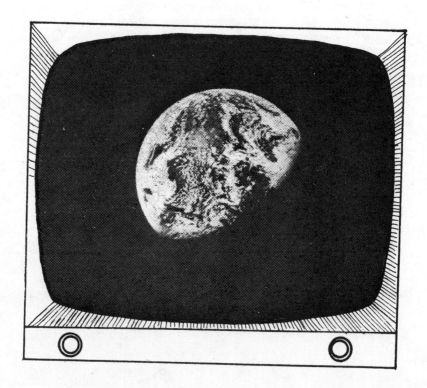

From the tops of mountains, from the gondolas of balloons, from the windows of airplanes and orbiting satellites, from the moon: we see and photograph our spaceship-planet, Earth. Through binoculars, telescopes, and radar spectroscopes we gaze out into our universe, and photograph it. With magnifying glasses, microscopes and electroscopes we peer in -- out to our universe -- and photograph it. With the expansion of our seeing has come the ability to record what we see for everyone to see. Satellite films of planetary weather patterns on Channel 2? Astronauts waving hello, live (!) from the moon on Channel 5? The inside of our bodies on Channel 12? Truly amazing! The invisible made visible at the flick of a switch! How much farther can our seeing go? There is no end in sight.

In order to see more and better, however, we must now focus our attention on two basic scientific assumptions which remain invisible until we "see" them. The whole structure of this, our world, as we have been taught to view it by our realistic science, is based upon an ideal thought-created mechanical abstraction -- the Ideal Eye. By international convention, we agree to use this Ideal Eye to establish standard measurements of the matter that surrounds us in empty space. Since physics' conception of space is bound by an absolute (the speed of light), we also measure time in standard units of linear progression (when only continuous cycles exist in nature). This Ideal Eye makes everything relative by splitting everything up and comparing this with that. It creates a world where there is no room for the Absolute. Finally, in order to quantify, analyze, and interpret these arbitrary measurements made through the Ideal Eye, we have created another ideal -- the observer of the objective universe, with a logical, rational, analytical, linear-sequential, thinking-computing mind -- an Ideal I. Consequently, we have been taught to localize our conscious self *in here* as opposed to what is not-self *out there*. This has led many of us to equate our existence with our body and our thinking mind. We see that we think, and therefore we think we exist as separate entities (to paraphrase René Descartes). Our ideal observer self, then, exists in an ideal space in an ideal universe, which it sees through an Ideal Eye with an ideally rational, logical mind. Does such a being exist anywhere?

These two ideals separate us from what is not us, like the separation of the eyes in our head. They not only separate us from one another but also alienate us from our real nature, our True Self. They have long dominated our scientific measurement, analysis, interpretation, and validation of information. Yet both of them ignore the basic facts of experience. No Ideal Eye exists in nature. Perception is relative to the observer's consciousness. No ideal objective observer exists either, since the validator of experience, the self, has been excluded entirely in the objective search for Truth, which ironically is supposed to be *self-*evident.

From these two ideal assumptions we have derived a framework of axioms, laws, and first principles based upon comparative knowledge, which is necessarily dualistic and, by definition, relative. The words of Einstein still remind us that:

> Pure logical thinking cannot yield us any knowledge of the empirical world: all knowledge of reality starts from experience and ends in it. Propositions arrived at by purely logical means are completely empty as regards reality.

We initially devised these two ideas for convenience, as necessary standards upon which to base certain agreements. Unfortunately, these standards have become the basis of how we are taught to view our world (from the standpoint of natural science) and, consequently, circumscribe our experience of it. They limit our possibilities for choice and potentials for growth to those allowed by "natural" law.

We will begin in Chapter Four with the Ideal I – the objective observer. More and more of us are finding that, in any investigation or experimentation we undertake, the *most important and relatively ignored variable* is the consciousness of the observer – our participating, experiencing self. Full recognition of the fundamental importance of consciousness brings the realization that any claim to objectivity which ignores consciousness is a sham. This results in a new role for us, that of participator. Since our thoughts, feelings, desires or states and levels of consciousness directly influence our perception, they must be taken into account and included in any observation we make or we are not being *completely* objective. This adds the experiential dimension to scientific experimentation. Theory is tested by practice which brings real results to us so that we can know the truth for ourself. We will become participators in a few experiential experiments to test the other ideal – the Ideal Eye of classical mechanics. By placing a mirror before our Ideal Eye, we will see if it sees as we have been taught that it does. We will find a whole new way of perceiving our universe and ourself -- one which turns our whole world view outside-in.

There are ways of "seeing" that go beyond our senses and rational mind. Sometimes called intuitive validity in science, these ways of knowing are merely an extension of our senses into the supersensitive realm of being -- intuition and imagination. An interesting thing is happening. Some of the finest minds among us are using the latest scientific tools to investigate how we interthink and interfeel across time and space. These ways of knowing are being empirically documented and visibly demonstrated. In Chapter Five, we will look into our inherent supersensitivity and latent potential for expanded seeing. The findings of the new sciences of parapsychology, psychotronics and psychophysics will be explored as they pertain to psi communication, direct communication via thought waves. Experiments are included so that we can test these abilities in ourself. Finally, in Chapter Six we will delve into the new world multi-mass-media communications systems which promise to extend our seeing to the furthest limits of the imagination -- and beyond!

SEEING THROUGH MATTER

THE CHANGING WORLD VIEW IN PHYSICS

The impact of modern physics on human society is extensive. The combination of natural and technical science continues to transform the world around us in both beneficial and detrimental ways. Its influence extends beyond technology to thought and culture where the concepts of physics help to mold our view of the world. But basic concepts in physics are changing.

> The concept of matter in subatomic physics, for example, is totally different from the idea of a material substance in classical physics. The same is true for concepts like space, time and cause and effect. These concepts, however, are fundamental to our outlook on the world around us and with their radical transformation our whole world view has begun to change.

The classical Cartesian-Euclidean-Newtonian idea system can no longer account for the phenomena which physicists and other scientists are seeing with their very own eyes. An entirely new world view is forming. To Dr. Fritjof Capra, research physicist in high-energy, subatomic physics (quoted above) this new world view is wholistic and essentially mystical, for it is conceptually close to Buddhist, Taoist, Hindu and Christian mystical cosmological teachings. It is based upon, but transcends, the quantum and relativity theories of Einstein. It is leading to the fundamental realization that mind and matter are not separate after all, but are one in consciousness. Modern physics is confirming the idea of Sir James Jeans that the universe is more like a great thought than a great machine.

The old world view was based primarily on three basic *idea-systems* -- Cartesian materialist philosophy, Euclidean geometry, and Newtonian physics. René Descartes divided the universe into two separate realms of mind (*res cognitans*) and matter (*res extensa*). This is the philosophic basis for the subjective-objective duality which has split science into many separate fields of investigation with no unifying principle. Three-dimensional Euclidean geometry has been taken as the scientific explanation for the true nature of space for over two thousand years. This geometry led Newton to his mechanics of three-dimensional flat space, with a separate dimension of time flowing in a straight line, from past through present into future.

The elegantly simple theories of Einstein have shaken physics and all of science from top to bottom. These theories have shaped new conceptions about the universe, ones which turn the old world view outside in. The old view of the universe as a great machine with unconnected and interchangeable parts led to manipulations which have given us many of the ecological problems we now face. These problems cannot be solved without an increase in awareness and an expansion of vision to a new level of thinking, where our mind is servant to the heart of Self rather than master of it. Einstein summed it up when he said: "the world we have made as the result of the thinking we have done thus far creates problems that we cannot solve at the same level we created them at."

RELATIVITY

Experiments with electricity and magnetism have led to the widely accepted axiom that "all electromagnetic waves (radio waves, light, X-rays, etc.) propagate through space at a very specific velocity, usually designated by the letter C (the constant speed of light, 186,000 miles per second)." (12) Einstein's special theory of relativity is based on the assumption that the speed of light is an absolute constant. That is, no matter how we measure it and whether we or the light source are moving, we should always arrive at the same fundamental velocity. After apparent proof of this was provided by experimental observation (the Michelson-Morley experiments) physicists reasoned that either the earth is standing still or there is something very wrong with Newtonian mechanics.

In relativity theory space is not three-dimensional and time is not an absolute, separate entity. Both are inseparably interconnected and form a four-dimensional continuum, space-time. We cannot reason about one without including the other. Furthermore, there is no universal flow of time as in the Newtonian model. The same events will be ordered differently by different observers moving at different velocities relative to those events. Therefore, all measurements involving space and time lose their absolute significance. Since our concepts of space and time are so basic to our understanding and description of the universe, modifying them means modifying the whole framework of scientific thought. We seem to be entering a space-time warp in world thought.

SUBATOMIC PHYSICS

Atoms, we have long been taught, constitute the basic building blocks of matter of which our body and our physical-objective world are composed. When physicists penetrated the submicroscopic world of the atom they found that Einstein's formula, $E=MC^2$, was correct — that matter is really a form of trapped light or energy. The tremendous light, heat, sound and wind which poured forth from the first atomic blasts boldly announced this fact to the world, and ushered in the "nuclear age" with a bang! It resulted from a *new view* of (or idea about) the atom, one that has not yet been fully realized in other branches of science. Rather than static, dead entities they (and all of matter) are very much alive.

Capra notes that:

Modern physics... pictures matter not at all as passive and inert, but as being in a continuous dancing and vibrating motion whose rhythmic patterns are determined by the molecular, atomic, and nuclear structures. This is also the way the Eastern mystics see the material world. They all emphasize that the universe has to be grasped dynamically, as it moves, vibrates, and dances: that nature is not a static, but a dynamic equilibrium. (3)

THE IDEAL I

Einstein's ideas have not only changed basic conceptions about space-time and matter, but they have energized physics to continue on into higher levels of awareness and thought. More and more scientists are becoming aware that the consciousness of the observer is an inextricable part of the process of scientific investigation. The ideal objective observer – the Ideal I – is the key hidden variable exposed by the new physics. According to physicist Werner Heisenberg all observations are relative and approximate, for "what we observe is not nature herself, but nature exposed to our method of questioning." (18) The design and arrangement of the measurement determine, to some extent, the properties of what is observed. "If the experimental arrangement is modified, the properties of the measured object will change in turn." By our very presence and our means of measurement we influence the properties of what we see. Apparently, this means that *whatever we believe to be true* at a certain level of observation *either is true or becomes true* for us. Nobel prize physicist Eugene Wigner states that "The being with a consciousness must have a different role . . . than the inanimate measuring device." (18) Physicist John Wheeler sees this issue as the most important feature in physics today and suggests we replace the idea observer with the idea participator.

> The vital act is the act of participation. "Participator" is the incontrovertible new concept given by quantum mechanics. It strikes down the term "observer" of classical theory, the man who stands behind the thick glass wall and watches what goes on without taking part. It can't be done, quantum mechanics says.

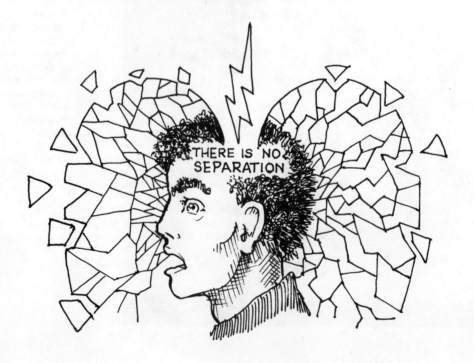

RADIATIONAL PARAPHYSICS: SUPERSENSONICS

Radiational paraphysics is a new-old science of vibrations in which the experiencer is an integral part of the experiment. It is based on our ability to sense vibrations through radiesthesia. Explained by Christopher Hills:

> Radiesthesia can be defined as the detection and recognition of nature's influence by the object under consideration. Recognition comes from the determination of the form or pattern of energy assumed by the object according to its structure. By the use of pendulums, rules and rods, special devices and witnesses, the operator is given selective powers which enable him to analyze a sample or discover an object underground. It can also be used for diagnosis in medicine or veterinary work, tracing the energy patterns in a living body. The detected influences are vibratory and radiesthesia's major fact is that a pair of substances of the same vibration are linked through space by harmonic resonance irrespective of whether solid matter is intervening between them. (10)*

Radiational paraphysics, sometimes called radionics, is the ancient art of divining brought up to date and refined into a participatory science. The most common form of this latent faculty in us is dowsing. Dowsers have been known to locate water, oil, mineral deposits, lost people, etc. when no one else could. In radiesthesia and radionics the detection of subtle vibrations is seen not as a psychic gift, but as a mental-biological sensitivity which about 80 percent of us have naturally, though generally unrecognized and undeveloped. Radiesthesia is covered in more detail in Chapter Ten.

The science of supersensonics is called by its originator, Christopher Hills, "the spiritual physics of all vibrations from zero to infinity." It is an inclusive term for the synthesis of the related fields of conventional physics, radiational paraphysics, radiesthesia and radionics -- which is the adaptation of radiesthesia to electronic tuning and instrumentation. According to Hills:

> To view these measuring devices as anything more than an extension of our senses is an error of perception. Einstein says all scientific measurements begin and end with sense experience. Supersensonics is directly concerned with understanding the perception of physical sense experiences and not ignoring the fact that it is always consciousness which ultimately makes sense of the senses. It also recognizes that it is always consciousness which must limit or enhance non-sensory perception. (10)

Supersensonics is fundamentally grounded in consciousness and relies on the operation of a series of natural laws. Beasley tells us that:

> These laws and the proof required to substantiate them involve the knowledge and techniques of conventional physics, as well as certain "para-physical" concepts which are not generally embraced by standard scientific approaches.... The human organism is "tuned" to the frequency of the incoming wavelength of the substance or the source being investigated. (2)

Through conscious development of supersensitivity, our nervous system can become sensitive to vibrations around us that extend far beyond our senses into the, so far, invisible, inaudible and nontactile. By holding in the hand or in the mind a sample or "witness" of a substance we wish to locate or divine the essence of, our consciousness will resonate with the vibrations of that substance. Through knowledge of the laws involved, we can then locate or divine that substance. Or in the case of diagnosis and treatment we can detect the characteristic vibrations of a disease, even at great distances from the person who is ill and send a counter vibration of healing energy (see Chapter Eleven on radionics).

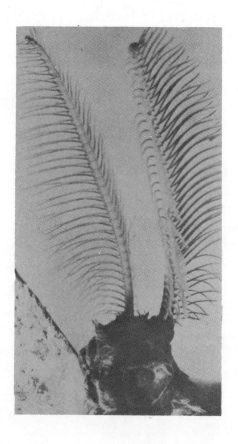

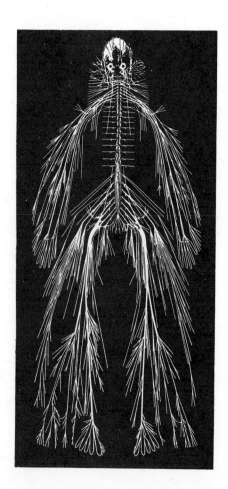

Considered biologically, Supersensonics is no more wonderful than other forms of perception by hand, eye, or ear. The radiesthesic faculty in man appears to be the same used by homing pigeons, dogs, and even moths, as a means of finding their way back home or of finding a mate. In fact, an examination of the antennae of insects reveals that one antenna is negative and the other positive and that they are used for vibratory wave selection within the insect's environment. It is since the advent of modern Supersensonics that we know that storks never nest above an underground stream, and that anthills are almost always over one. (10)

The antennae of the moth (left) are mirrored in our body in the supersensitive nervous system (right).

THE OPEN MIND

The nervous system of our body (the tree of knowledge) and the nerve dendrites of our brain (the tree of life) form a feedback network for sensing all vibrations, from zero to infinity. The idea that we can consciously use our body as a detecting and measuring instrument turns conventional science upside-down. But supersensonics meets the basic criterion of New Age science – it includes the consciousness of the participator. It is proving to be a discipline (in the true sense of the word) which is helping to heal many of the self-created divisions which separate us from our Self. By focusing on the consciousness of the participator, supersensonics places a mirror before the Ideal I which has long surveyed the universe and divided it up according to arbitrarily chosen fixed standards. By helping us develop intuition it turns our attention and awareness to our feelings and challenges us to reexamine ourself and expand our awareness. It reminds us that the best mind is one that is open to the center or heart of our being.

The branch of science that studies how we see is called optics. It is here that the Ideal Eye of classical mechanics meets the Ideal I of the objective observer in a mirror of the Self.

CYCLOPS: THE IDEAL EYE

There is a huge cyclops eye sweeping our world today. By international convention we have created an Ideal Eye of constant characteristics which responds mechanically to light radiations of different wavelengths emanating or reflected from illuminated objects. It peers into three-dimensional, flat, empty space and through one separate dimension of time. It is this Ideal Eye which is likened by physicists and opticians to that of a camera lens with set exposure times.

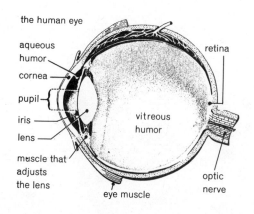

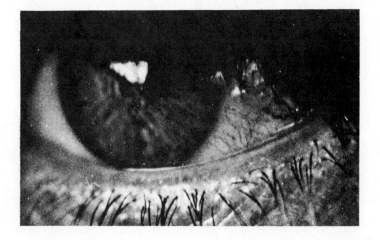

However, because of the new world view mentioned earlier, we now know that space-time is an inseparably unified four-dimensional continuum in which matter is more realistically viewed as waves of vibrating energy patterns. These pulsating waves of rhythmic, dancing sounds and colors fill all of space, connecting everything and everyone. But the Ideal Eye cannot perceive such a world since its vision is confined to the purely physical world of solid, static, separated objects. Its vision is frozen by the ideal objective observer -- the Ideal I -- which has been educated into us to the detriment of our experiencing, participating self.

If we break the spell of the Ideal I new ways, super ways, of seeing our world and ourself become available. To open ourself to these new ways we must first confront the Ideal Eye with itself by placing a mirror before it. That mirror is us looking at and through the Ideal Eye, exposing it to the "mirror of collective reflection," and finding out for ourself if it actually sees as classical mechanics has taught us that it does. And here, ironically, is where we confront the big question of Self in scientific thought. As Christopher Hills reminds us:

> By assuming that light as seen by the living eye internally has the same effects as do external radiations on an inert sensor, a contradiction has arisen in the very results of science itself... everyone knows that the physiological process which leads to the experience of light is experienced and ultimately interpreted by consciousness -- the Self. Yet that Self, which is the validator of the experience, has been ruled out totally in the objective search for Truth, which is supposed to be self-evident. (10)

The question of Self was filed away under the "subjective" science, psychology. For ages no one has been looking through the Ideal Eye, yet all measurements are supposedly made by and referred to this eye. The physical explanations of vision have long been assumed, without question, to be mathematically and geometrically sound approximations. What was a purely physical phenomenon in the mechanics of classical optics is now, with us as participators looking at and through the Ideal Eye with our own eyes, a psychophysical phenomenon. That is, we unite the physical seer (the eye of optics) with the psychic seer (our consciously participating self). Eyesight and mindsight become insight when we see how we really see.

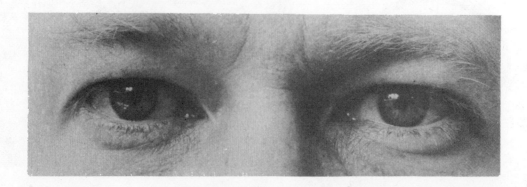

HOW WE SEE: PSYCHOPTICS

Professor Vasco Ronchi of the National Institute of Optics, Arcetri, Italy has made a detailed study of the physics and psychophysiology of seeing. The results lead him to conclude that all information received through the senses contributes "to the formation of the apparent world, which is essentially psychical, subjective, internal to the observer, and partly, even if not wholly, hard to be put in objective terms. Thus, the apparent world is an ensemble of psychic representations." He points to the "hidden hypothesis" of which many of us are unaware. This simply means that we ordinarily assume that everyone sees, or experiences through the senses, in the same way. He points out how this attitude has even crept into our use of language and has come to dominate our thinking:

> Several quarrels would be avoided if, instead of asserting, as is often done, "you said this," one said, more accurately: "I hear you saying this." This appears as a shade of meaning, but the difference between the two is enormous. (10)

Ronchi has performed detailed, systematic and exhaustive research into the process of seeing.* By including the experiential confirmation of a participator (the seer behind the eye) and asking us what we see, he has found that the classical explanations of seeing are, in most cases, totally and erroneously false. In fact, Ronchi points out that many of the past experiments appear to have been rigged to prove a particular hypothesis which holds true only within very limited parameters. If our eyes move from the required spot, or if we direct them somewhere else, the classical physical explanations immediately fall apart.

> The dangerous behavior, which is scientifically unacceptable, consists in hiding the basic assumptions and in using tricks to prove as true the theoretical conclusions, by carrying out tests in very particular conditions, accurately chosen in such a way that the hidden assumptions be valid. But, in this case, the study is not a scientific activity, it is a trick. (10)

Ronchi has performed many experiments with converging and diverging lenses, flat, concave and convex mirrors, and the real, naked eye itself. He notes that in the case of converging and diverging lenses, "the disagreement between classical formulas and experience could not be wider than it is." The principle of magnification is completely unexplainable by classical formulas.

---

* This research is covered in detail in Volume III, "Supersensonics." Many of Ronchi's experiments, with diagrams, descriptions, and explanations are covered therein.

It seems natural to conclude that the classical theory is unable to answer the initial questions: it cannot explain why one sees enlarged figures through a magnifying lens; it cannot say how much such a lens magnifies. The classical theory either limits itself to vague reasonings, derived through a purely mathematical procedure from the basic formula, or withdraws to conventions, which cannot solve any problem in agreement with the reality. (10)

This is especially true in the case of virtual images.* Virtual images are what we see in mirrors, under magnifying glasses, in hot deserts under the right conditions (mirages), etc. It is rather ironic, but quite fitting, that the Ideal Eye should fail so miserably in explaining the image of itself in a mirror. Confronted with itself, it cannot believe its eye! According to classical theory light rays take two paths when they strike a mirror. The "actual" path is called "real" and when light rays take this path from a mirror they are reflected at an angle equal to the angle of incidence. Other rays supposedly take another path, expressed as a mathematical representation, and form an image behind the mirror in reversed symmetrical orientation. However, as Ronchi points out, "behind the mirror is physically nothing of what we are concerned with. The rays are exclusively in front of the mirror." Consequently, the light rays cannot really go behind the mirror and the image is not real. It is called a virtual image.

"But the very meaning of the word virtual is not given, and one limits himself to saying that this term is used to indicate the images which are on the back-prolongation of the rays emerging from the rear of an optical system in general, and to distinguish them from the images that are on the actual path of the rays and which are called real. Why? No explanation is given. They are terms fallen from the sky, and as such they are accepted, having only a shadow of a meaning, of a purely geometrical nature. . . . . . .

. . .it is the observer who sends the effigy behind the mirror. This is a purely psychical entity which exists only if there is an observer who created it and localized it in the position that is required by the rule of the telemetric triangle." (10)

What we see, Ronchi concludes, is a series of images or psychic representations built up in our minds. First, there is a physical phenomenon which he describes as an atom cloud, which emits radiant energy.

---

* A virtual image cannot be shown on a screen. It can only be seen by an eye. See Chapter Nine on images and the creative imagination.

> The emitted energy propagates in the surrounding space, crosses other atomic clouds with shape and composition different from that of the surrounding medium, is deviated by those clouds and is concentrated in an orderly distribution, which is in some relation to that which it had been emitted by the atoms of the object. This new orderly distribution is called ethereal image and is a phenomenon of physical nature. (10).

The radiation from the ethereal image enters the eye and impinges on the retina, forming the "retinal image."

> There follows a physio-psychological process which results in the localization outside the eye of a luminous and colored effigy, whose angular aperture is equal to that of the ethereal image as seen from the eye. This image is the optical image and has a psychical nature. (10)

In short, vibratory wave patterns come through our senses and interact with wave patterns in our mind. The two combine to produce a picture of the world that is *apparently* real, solid, and objective to us the observer. This picture appears to exist "outside" our body, when really it is inside us, inside our consciousness. Perception, like space, time and matter, is relative. Dreams are like that. The alarm clock is ringing. Time to wake up. Do we become participators or remain observers?

OUR MIND'S EYE

The basic concept in psychoptics is that all perceptual processes are psychical entities, representations of the so-called "external-objective" world that are built-up in the mind. The mystical tradition warns that sensory perceptions are phantom illusions, the dance of Maya. Christopher Hills reminds us that:

> The whole Adam and Eve allegory is an outline of the process by which the "tree of knowledge", the nervous system, tempts us to accept sensory knowledge as real when all it does is present comparisons. Comparative knowledge being dualistic and not direct perception, was called knowledge of Good and Evil. Whereas the "Tree of Life", which made one immortal, was possible through the direct perception faculties of the nerve dendrites, brain pathways and trunkways of the neo cortex ... If we look at our own nervous system and the brain as a root system sucking in nourishment of "prana" or life force into an energy system, we begin to see that the experience we call ordinary waking "consciousness" is merely the imaginative formulation of concepts and does not effectively lead to reality at all. (10)

We can easily confirm or deny the classical theories by participating. Ronchi provides several experiential experiments to prove that what we sense is inside us. A few of these follow:

The simplest of these observations can be made by pressing an eye with a finger, in such a way that the eyeball rotates a little within the orbit: the apparent world splits at once, and one of the two copies rotates concordantly with the rotating eyeball. Clearly, the objects are not split nor in motion. Thus everyone should ask the question: what is the second figure, the one which rotates together with the eye?

Next, if one rotates one eye on a side and the other eye on the other side, both the copies of the apparent world move: each one moves concordantly with the corresponding eye, and nothing remains there where the observer said that he was "seeing the object", before rotating the eyes.

By rotating both our eyes, both copies of our apparent world move. Ronchi comments:

> This phenomenon, so common and so simple, seems to be explained only by considering the external world as an effigy, related to the eyes. Indeed, one should rather say that the apparent world which one sees when he uses both eyes, without any pressure on them is an effigy created by the cooperation of both eyes, and that when this cooperation is somehow altered, each eye creates its own effigy. (10)

He offers another example:

> If one looks at a watch placed on a table at a distance of about 30cm from the eyes, and at the same time he introduces a pen about midway between the watch and the nose, he cannot see either one pen or one watch, but the cases are two: either a watch is seen between two pens, or a pen between two watches. And still one knows very well there is only one watch and only one pen. (10)

There are several other examples, but we will end here with one more. A basic assumption of classical mechanics, and the basis of perspective that is taught in schools everywhere, is the rule of the telemetric triangle. Briefly stated, this says that the dimensions of an object viewed are directly related to the angle it subtends at the observer's eye. If we place our hands in front of our eyes at distances of twenty and forty centimeters respectively, we can clearly see that the farther hand is not much smaller than the nearer one. But according to the rule of the telemetric triangle, the angle subtended by the nearer hand is twice the number of degrees as the angle subtended by the farther one. The ratio, then, is two to one, and the farther hand should be half the size. *But it's not.* These elegantly simple and obvious participatory experiments reveal the perils of being an objective observer and allowing ourself to agree with everything without taking the time to participate in checking it out to see if it really works.

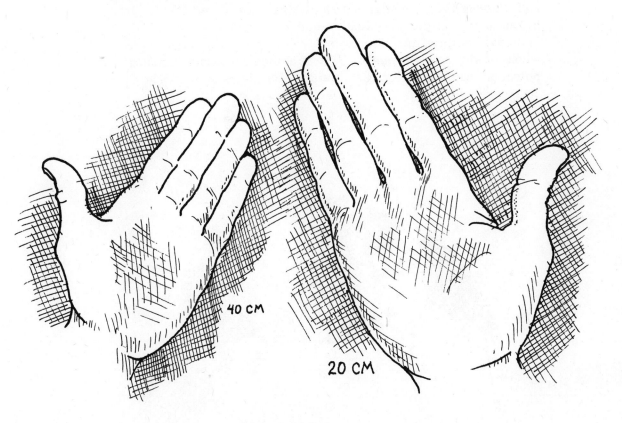

OUR MENTAL BLINDERS

These comments about seeing, Ronchi notes, apply to the other sensory processes as well. The energy field radiational waves (sensory stimuli) are effigies which only exist in an apparent world and arrive more or less deformed by our minds. This occurs because our mentally stored values, beliefs and symbol systems all conspire to cause us to screen incoming information through various conceptual models or self-imposed ideas of the world (world views). Mathematical, geometrical, intellectual, etc. concepts filter or "veil" our perception of reality. Notes Christopher Hills:

> A conceptual model is not a mock-up or a real image and in fact may not be concrete or visualizable in real terms. However, the image or patterns of information in a computer or the intangible set of mathematical relationships in the mind, is a product of our imagination and as such is a determinant of our reality, whether reality is considered virtual or actual. (10)

Is everything we see really constructed from interfering wavefields of thought taking place in our consciousness? In other words, do we really look through eyes at matter and empty space, or is everything an image in our mind? Is the "external" world of the observer becoming the "internal" world of the participator? This perspective literally turns our whole world view *outside-in*. This is a thought-warp of tremendous potential significance for the future of science, and for each of us.

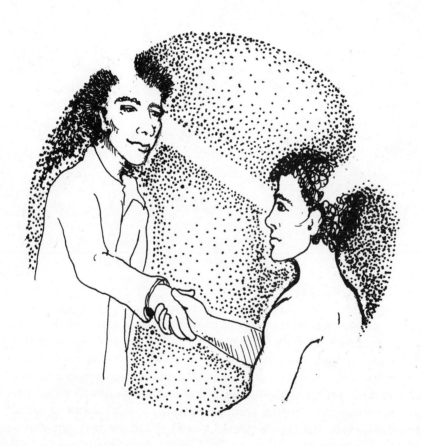

EXPERIMENT SECTION FOR CHAPTER FOUR

MIRROR GAZING

Many of us automatically focus our eyes at the surfaces of things, instead of seeing deeper into their true nature. This unconscious habit arises from our view of ourself as a being occupying a solid material body that is separate from other bodies. However, as we learned in Chapter Two, the physical view of the world is only one of seven levels of consciousness, and matter is really patterns of energy or crystallized light. By practicing techniques which allow us to unfocus our eyes, we gradually release our mind from its attachment to the physical and prepare ourself for experience of other levels of reality. In this way, we open ourself to expanded seeing of the more subtle surfaces and interiors of things, which we usually do not see because we have not trained ourself to see them. The aura is a good example. Just because we cannot see it does not mean it is not there. We cannot see X-rays either, yet they take strange pictures of the insides of things. In fact, X-ray vision may be possible if we learn to focus our eyes and our attention within someone rather than merely looking at the surface. Mirror gazing is an excellent way to train the eyes for these purposes.

STEP ONE: Sit in front of a vertical mirror, about six inches away. Close your eyes, breathe deeply and concentrate your attention on the brow center. Now open your eyes and look at the spot on your forehead which roughly corresponds with that interior center. Focus your eyes on that spot. Now try to look into each eye with each eye, first separately and then together so that they are parallel. Notice how, when you think of something such as a distant scene or an idea, your eyes change with the change in thought.

STEP TWO: Hold a piece of cardboard (about six inches square) vertically between your nose and the mirror. Now concentrate on seeing only one eye, your left. With practice, you will be able to look fully through the left eye with all your consciousness and see only the left eye in the mirror. And yet you know full well that both eyes are open and you should see both eyes. As you think about your right eye, your consciousness will act through the eyebeam to make it reappear.

STEP THREE: While still holding the cardboard between your face and the mirror, your eyes may begin to "float" in front of you. If you make the self-suggestion they they will both float toward the middle, gradually they will begin to merge until you see only one eye. This is the eyebeam concentrated into one single coherent ray of penetrating light which can penetrate to the deepest hidden truths. When the consciousness which sees is pure, then the eyebeam is capable of seeing through matter like an X-ray, across space like a huge telescope, into other dimensions of time, and deep into the heart of every living thing.

SEEING THROUGH MIND

PSI COMMUNICATION

Psi, the twenty-third letter of the Greek alphabet, means mind. Psi communication, then, is direct information exchange via thought. It is the psychotronic counterpart of our electronic mass media communications systems (explored more fully in the next chapter). As noted in Volume 1, Soviet scientists are reported to have recently discovered a brain-generated frequency of one hundred million (100,000,000) cycles per second, called the "ultra-theta." Vibrations at this rate, in excess of television waves, are capable of extending around the world. This provides a physical explanation for telepathic communication from brain to brain, over extremely long distances. Professor I.M. Kogan, physicist at Moscow's Society of Radio, Electronics, and Biocommunication, has carried out studies with two of the Soviet Union's star telepaths, biophysicist Uri Kamensky and actor Karl Nikolaev. (17) Both Kamensky, who was in Leningrad, and Nikolaev, who was in Moscow, were monitored simultaneously with electrophysiological equipment. This recorded brain waves, heart beats and other physical parameters of both men. Both had practiced yogic breathing exercises and visualization techniques to "tune in" to each other. They devised a special code of dots and dashes by which letters were to be sent. They successfully communicated over a distance of several hundred miles a four-letter word, "Ivan." This experiment may prove as important as Alexander Graham Bell's "Come here, Watson, I need you," for it indicates that there is the equivalent of a cosmic television screen in our heads which brings information about things that are happening somewhere else, if we have it "switched on."

Even faster vibrating, yet-to-be discovered frequencies emanate from our entire being beyond the ultra-theta. Only the instrumentation to detect them is needed before we will accept them as real. Well, as we will soon see, that instrumentation already exists. Nature has provided it right here inside our body. Our supersensitive nervous system is "crowned" with a huge network of billions of nerve dendrites in our brain. Besides the nervous system, every cell in our body is interconnected by the microtubular waveguides (explained later) of the

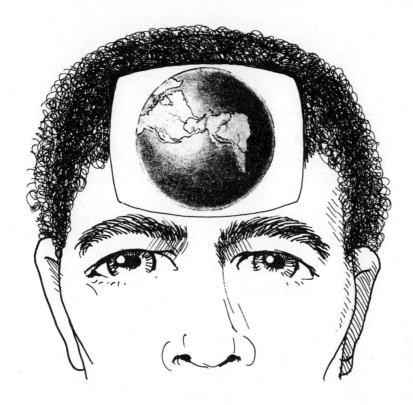

acupuncture meridian system. Information is stored in the DNA of every cell in our body and throughout the nerve networks. The now familiar alpha, beta and delta brain waves are just the lower range octaves of a huge spectrum of mind-generated radiations. Thought waves are real. Our thoughts create our experience. We live in a thought-formed world. (As we will see in the radiesthesia section, Chapter Ten, thought waves can be detected through supersensonic techniques.)

PARAPSYCHOLOGY

Para, in Greek, means beyond. Parapsychology is the branch of science devoted to the study of experiences, visions and insights that take place beyond sensory perception, beyond time and space. The main areas of study are extrasensory perception (ESP) and psychokinesis (PK) although other related phenomena are also considered. ESP is generally divided into three main areas of inquiry:

> Telepathy - Our ability to communicate mind-to-mind, to "read" each other's thoughts.
> Clairvoyance - Our ability to experience *across space* directly without contacting another mind.
> Precognition/Retrocognition - Our ability to experience *across time* into the future or into the past.
> PK is our ability to move or otherwise influence "matter" directly, through concentrated thought alone.

Dr. Rammurti S. Mishra, who besides being a swami in the yoga tradition is also an endocrinologist, neurosurgeon, psychiatrist and linguistic scholar, warns us that:

> Supernatural powers occupy the central place in psychic research. In Yoga, the central feature is "Nirvana Kaivalyam" (absolute liberation), where miracles and magical powers are regarded as obstacles in the path of self-realization although they are by-products of Samadhi, Cosmic Consciousness. Yogins themselves are reticent about these "miracles" although they are acquainted with them. (16)

The reason for this reticence according to Christopher Hills is that one can have all these psychic powers and be just as egotistical and selfish as anyone else and therefore they do not indicate spirituality.

EXTRASENSORY PERCEPTION

ESP, once generally thought to be superstitious hocus-pocus or black magic, is now viewed scientifically as a communication process that can be studied empirically. So far, no reliable relationships have been found between ESP test scores and sex, age, blindness, mental or physical illness, or intelligence. Scores have generally been better when methods of testing included rewards, novel test situations and feedback on scoring results. This has led to study of the effects of emotional, feeling states on scoring. Here, parapsychologists have uncovered what they call the "experimenter effect." One study found that whether the experimenters smiled or not or whether they greeted their participators in a friendly or cold manner had a noticeable effect on ESP scores. This led one parapsychologist to note that "the parapsychological experimenter effect has revolutionary implications for normal psychological research." Another interesting finding is that higher scores for receiving are usually made by those who are relaxed or in trance or hypnotic states. In other words, altered states of consciousness seem to accompany experiences of extrasensory perception. Finally, investigators have discovered the phenomenon of "psi missing," in which participators score far lower than chance expectations. This indicates that there is a psychological state (doubt) in which psychic material is repressed from consciousness. A series of tests with high school students in India found the following personality traits associated with high and low ESP test scores. (16)

Positive ESP Scores	Negative ESP Scores
Warm, sociable	Tense
good-natured, easy going	excitable
assertive, self-assured	frustrated
tough	demanding
enthusiastic	impatient
talkative	dependent
cheerful	sensitive
quick, alert	timid

Positive ESP Scores	Negative ESP Scores
adventuresome, impulsive	threat-sensitive
emotional	shy
carefree	withdrawn, submissive
realistic, practical	suspicious
relaxed	depression-prone
composed	

PSYCHOKINESIS

Drs. J.B. and Louisa Rhine conducted rigorous, carefully controlled parapsychology experiments at Duke University in the decade of the 40's. They wanted to satisfy science's demand for a rigid experimental procedure, with controls, double blinds and statistical analysis. This involved the throwing of dice by mechanical means, so that no one ever touched the dice. Most of the participators were students at the university who volunteered to try to influence the dice mentally. The number of correct calls out of thousands of throws was very high. The probability against such a high score by chance was an astronomical 10^{115} (10 with 115 zeros after it) to 1. (23)

More recent experiments with PK reveal another variation of the experimenter effect, this time with machines rather than people. Buried deep beneath the Stanford Research Institute lies a sensitive magnetometer shielded by a superconducting electromagnet. At the time of the experiments it was being used to measure the rate of radioactive decay of isotopes. This instrument has a special casing which screens out all known radiations, so that no signals can disturb it from the "outside." And yet Ingo Swann, a gifted psychic, by concentrating his attention inside the device, repeatedly was able to deflect a recording needle inside the magnetometer, doubling the recording output for about thirty seconds. Other unusual disturbances were also noted by participating physicist Dr. Harold Putoff. (16)

Another experiment involved psychic healer, Dr. Olga Worrall. She changed the environment in a cloud chamber (instrument used to detect sub-atomic particles) simply by placing her hands near it. Pulsating waves were seen in the chamber moving parallel to her hands. When she shifted them ninety degrees, the waves also slowly shifted until they were again parallel with her hands. Any attempt to explain this was complicated by another experiment. The physicists involved called Dr. Worrall by phone as she sat in her home 500 miles away from the chamber. Again, the same pulsating waves formed inside the instrument and persisted for approximately three minutes. A skeptical physicist who was present noted that this would only be significant if repeated. They called her back and this time the waves not only reappeared but lasted for eight minutes!

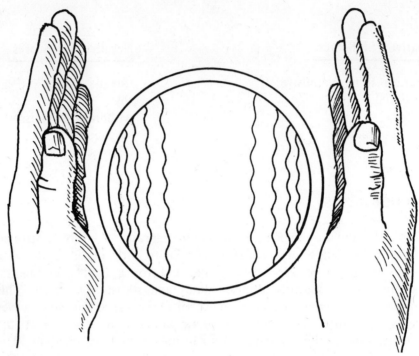

Then there is Uri Geller. This young Israeli is confounding not only scientists but television audiences as well. In different PK experiments he is reported to have bent and broken metal objects, erased magnetic tape, made things disappear and reappear elsewhere, caused the hands of a clock to move, altered a magnetic field measured by a magnetometer and altered the decay rate of radioactive isotopes. One scientist is reported to have unexpectedly handed Geller a bean sprout and asked him to "make time run backwards." (16) Geller closed his hand over the sprout. Thirty seconds later, he opened it. In place of the sprout was a whole, solid mung bean! If this experiment can be repeated and verified it indicates a psychokinetic influence involving organic time.

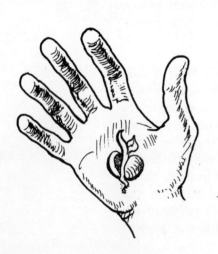  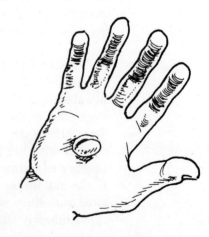

The most striking thing about Geller, however, is his popularity with and effects upon TV and radio audiences. A Berkeley research group conducted a follow-up study of the reactions of some of those who witnessed Geller's performance on TV. Many people reported the experience of unusual or telepathic phenomena. Some of them were even able to produce psychokinetic effects themselves. Parapsychologist Dr. Jeffry Mishlove reports that "on occasions when I have broadcast radio interviews with Uri, dozens of listeners have reported psychokinetic phenomena in their own homes." (16) Geller appears to be popular elsewhere, too. He has appeared on T.V. in several countries where thousands of people, especially children, are reported to have bent spoons after seeing him perform. Is this a form of mass hypnosis? Or is it the Alice in Wonderland "You can if you think you can?"

The above examples of our ability to influence scientific instrumentation by our thoughts point to the overriding importance of our attitude and states of consciousness as participating researchers. They illustrate the necessity for purity of motive in research, since we may be creating whatever it is we are looking for. In addition, it is highly possible that our instruments do not give a truly objective description of what we are measuring, since our will may interfere. These examples indicate the strength of concentrated awareness when focused and consciously projected.

For far too long our educational systems have been dedicated to logical-rational mental development, with the sciences receiving much more financial support and active encouragement than the arts. Now that the great myth of the objective observer is exposed, however, many of us are questioning what we have been taught. If some of us can travel the universe of the mind, then potentially all of us can. If concentrated thought can move or change physical objects, then what is "matter" but thought?

More and more of us are becoming participators in intuitive and imaginative explorations of consciousness. Millions of us have experimented with psychoactive drugs such as LSD, mescaline, marijuana etc. From these experiences we know that perception is relative to our state of consciousness as perceiver. Having had a tantalizing foretaste of the delights of expanded consciousness, we have at the same time learned the necessity for purifying consciousness before expanding it. We cannot expand what we do not have. If we do not have love in our heart, how can we expand it to include all? No matter how high any drug-induced experience may be it must eventually end and we are right back where we started. In addition, a "bum trip" can be dangerous for our sanity. These considerations have drawn unknown millions of us to meditation in search of that blissful state of eternal grace which bestows absolute peace. We are helped along the way by tuning in to that still, intuitive faculty in us which sees where eyes of flesh cannot. Our crown chakra is linked directly with that universal field of vibration (ether) by which everything is interconnected and in which everything has being. Through understanding how the mind works intuitively and imaginatively we learn how telepathy works. From telepathy spring the other psychic powers.

BREATHING AND TELEPATHY

Recent scientific research is revealing that breathing negatively ionized air brings beneficial effects such as lowered blood pressure, slower, deeper respiration and a general feeling of mental alertness and well-being. (2) It also increases telepathic receptivity by inducing a state of relaxation. In physiology this is called cholinergia. It is a state of receptivity which is induced through relaxation techniques such as hypnosis, breathing exercises,etc. This occurs when the parasympathetic nervous system acts on effector cells (muscle and gland cells) to liberate acetylcholine. This, in turn, brings a relaxation response in our body which is reflected in a synchronous brain wave pattern.

The opposite state is excitation, called adrenergia. It is characterized by acute alertness. It is difficult to induce artificially except when breathing positively ionized air, but it arises naturally in stressful situations. This occurs when the sympathetic nervous system becomes agitated through the release of adrenaline and adrenaline-like compounds in our body.

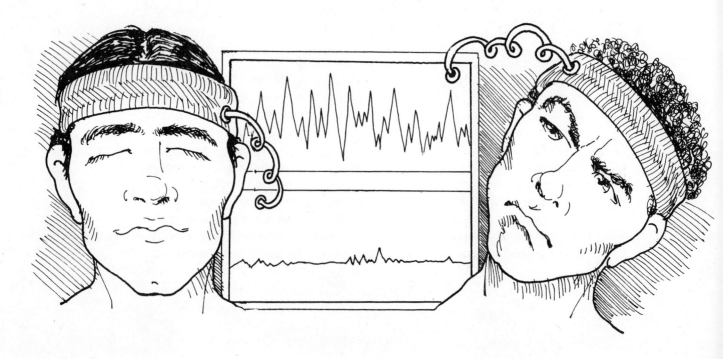

Experimental testing shows that a telepathic sender is most effective when in a state of tension (adrenergia). Medical physicist and parapsychologist Dr. Andrija Puharich found in experiments with psychic Peter Hurkos that Hurkos was afraid of electricity. Hurkos' ability to send telepathic messages increased considerably when he was asked to sit with his foot on a plate carrying a 10,000 volt direct current charge. (22)

Other experiments have shown that a telepathic receiver is most effective when in a state of relaxation (cholinergia). Psychic Eileen Garrett is reported to have been most successful when breathing air with an excess of negative ions. Puharich has found that the highest telepathic scores for receivers are directly associated with left nostril (receptive, negative, lunar, Shakti) breathing. As we found earlier, left nostril breathing ionizes oxygen atoms negatively (adds electrons). This helps explain Puharich's success. So if we wish to become more receptive to telepathic messages we can practice left nostril breathing. If we wish to become better senders we can do right nostril breathing. We don't need a laboratory full of instruments to find out about telepathy. We have all the necessary ingredients right here inside our consciousness.

TELEPATHY AND THE WEATHER

Stormy weather has long had a reputation for interfering with telepathic abilities. Poor weather conditions have been correlated with lower negative ion concentrations in the air, because water molecules tend to pick up the free electrons and negative ions. The ionizing effects of different weather conditions can be duplicated by using single and double Faraday cages (similar to preparations for practicing advanced dowsing experiments explained in Volume III). These ionic effects on telepathy indicate the presence of some kind of mediating field for reorganizing energy patterns into forms that are meaningful to us.

PSI PLASMA

Psi means mind. Plasma, as we have seen, is used to describe a collection of electrons and ions which are shaped into a form by magnetic fields and which maintain that identity. Puharich uses the concept of psi plasma in the original Greek sense of form. (22) Form is the common factor of both "internal-subjective" and "external-objective" experience. Any communication, electronic, psychotronic or otherwise, requires the medium of form for capturing and conveying meaning, thus connecting the two ends of any communications link. Psi plasma is molded by our thoughts, our mental force-fields. It acts upon bioplasma to produce extrasensory effects which transcend our normal biological senses. Psi plasma is a means by which we can experience, if we wish, beyond our body as a mobile center of consciousness (MCC). This nuclear mental entity has all the characteristics that we normally attribute to mind: perception, feeling, association, reasoning, memory and even the creation of ideas.

The most important factor in telepathy, according to Puharich, is the relationship between the biogravitational force-fields of the sender and the receiver (we will explore the concept of biogravity later). That is, if our biogravitational forcefield is strong enough we will be able to attract the attention of a receiver. (Or in the case of clairvoyance and precognition we attract an event across space and time.) When these psi plasma fields meet between two or more of us there is a sharing of being, an exchange of information and a flash of imagination or intuitive insight -- an imprint -- in the consciousness of the receiver. This connection takes place beyond the confines of space, time or intervening "matter." The speed of conscious thought is faster than the speed of light. In fact, time has no meaning at the highest levels of consciousness where all events exist concurrently in the eternal now. Now is more than the immediate now for a being experiencing the highest states of consciousness. We only experience a space-limited, time-bound world when we separate ourself from that great Self which is All.

> According to Yoga, the Self of man is omnipresent, omnipotent, and omniscient although such power is not fully manifested in an untrained mind. Individual mind is representative of Cosmic Mind, so it can communicate with any other individual mind or phenomenon anywhere, in any planet and solar system. (16)
>
> Rammurti S. Mishra, M.D.

PSYCHOTRONICS

In the early 1960's, in order to synthesize serious research efforts into the interconnections of energy, matter and consciousness, several scientists from many countries created the science of psychotronics. This is an "umbrella" science which combines all pertinent research into the effects of subtle energies and forces not explainable by the laws of physics. Psychotronics is formed from two Greek words: *psyche* meaning breath or spirit and *tron* meaning instrument. The basic idea revolves around the concept of a carrier wave for thought energy, the psychotron. This is a theoretical particle wave which vibrates beyond the speed of light, up to the speed of consciousness (instantaneous). This energy is found to supersede the more familiar physical forces of electromagnetic, nuclear and gravitational waves because it can counteract them. A specially designed psychotronic "accumulator" charged with thought energy can attract nonmetallic particles to it like a magnet. It can also pull iron filings away from a strong electromagnet. This indicates the presence of a stronger attractive force than magnetism -- possibly a biogravitational force. It becomes "lower" forms of energy and even "matter" by following the laws of transformation. That is, it gradually slows down its rate of vibration as it meets other waves, in the same way that the gas, steam, changes to the liquid, water, and to the solid, ice, as temperature decreases.

In June, 1973, distinguished scientists from twenty countries met in Prague, Czechoslovakia, for the First International Conference in Psychotronics. This was a major interdisciplinary gathering, for in attendance were physicists, chemists, electronics specialists, biologists, psychologists, psychiatrists, etc. Their purpose was to officially establish the science of psychotronics: "the study of all interactions between man and objects both animate and inanimate." This includes the highly unconventional study of our interaction with machines.

> If the consciousness of man influences the nature of the world around him, then it may also influence the instruments with which he works. Perhaps, then, our instruments do not give a truly objective description of what we are measuring. (17)

Dr. John Jungerman, nuclear physicist at the University of California at Davis (quoted above), described his work with the laser source interferometer, which is sensitive to minute forces of only a few micrograms (millionths of a gram). Some persons are able to influence this delicate device simply by their very presence.

Czech engineer Robert Pavlita went one step further. He demonstrated his psychotronic generators. These are metal objects which he has designed in various shapes. The shape determines how the energy is used to perform different kinds of work. Pavlita charges these devices with his own thought energy. Placing a charged generator near a jar of muddy water speeds the precipitation rate. Putting one next to a plant increases its growth rate. And so on. His daughter placed a generator against her right temple, then placed it near a device resembling a windmill. The device revolved in a clockwise direction. Next she placed it against her left temple, returned it, and the windmill type wheel spun in the opposite direction. This indicates that the polarity of the brain causes psychotronic energy to be polarized. Members of the audience at the conference were invited to try to duplicate the experiment. A few of them succeeded.

Then came another Czech engineer, Julius Krmssky. He demonstrated how energy emanations from his hands can cause objects floating on water to move forward and backward. Next, he made an object suspended by a string in a bell jar (described in the experiment section) rotate clockwise and counter clockwise by rotating his hand near the jar. Finally a film was shown in which Krmssky was able to turn on a light by focusing his eyes on a diode switch.

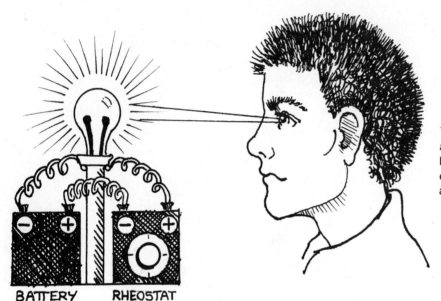

You can do it too! Set the voltage adjuster at just below the trigger voltage of the diode bulb. Project bioplasma (consciousness) out of your eyes through concentrated attention and watch it light up.

This feat points to an energy coming out of our eyes, which is called the eyebeam in supersensonics and is used to amplify the diviner's reactions. This same experiment was performed for many physicists by Christopher Hills in Jamaica in 1963, in London between 1966 and 1970 and for Dr. Andrija Puharich in New York in 1967. Soviet scientists, using a special film emulsion sensitive to the ultraviolet spectrum along with selective filters, have recorded some very strange pictures generated by the eyebeam. Inyushin reports that "we discovered an emanation from the eyes of animals and humans." Opaque screens and thin metallic sheets were placed between eyes and film and still "we obtained from human eyes, under conditions of auto-suggestion, after exposure of only one-thousandth of a second, very clear images on the emulsion." (19)

These demonstrations signal the advent of a new scientific revolution. The implications of widespread practical application of psychotronics can only be guessed. But one thing is certain, society as we know it will be transformed. We need only look at the profound worldwide transformations caused by electronics (explored in the next chapter) to get an idea of what psychotronics and supersensonics will bring. Of course, all this remains only theory unless we prove to ourself that psychotronic energy really works. Experiments are included at the end of this chapter.

Preliminary research indicates that there are at least two types of generators that harness psychotronic energy. The first, as we have seen above, is the bioenergy generator. It requires the conscious focusing of our bioenergetic field to make it work. The second type is called the cosmic generator. It works automatically. Where bioenergy devices perform relatively simple tasks, the cosmic ones have a range that is virtually unlimited (the limitations being set by the consciousness of the operator). One such cosmic psychotronic generator is the Great Pyramid of Cheops in Gizeh, Egypt -- the seventh and last remaining of the seven wonders of the ancient world. It has been discovered that the King's chamber was specially designed to select out one particular ray of concentrated life energy, the Pi-ray, and focus it in the King's coffer.

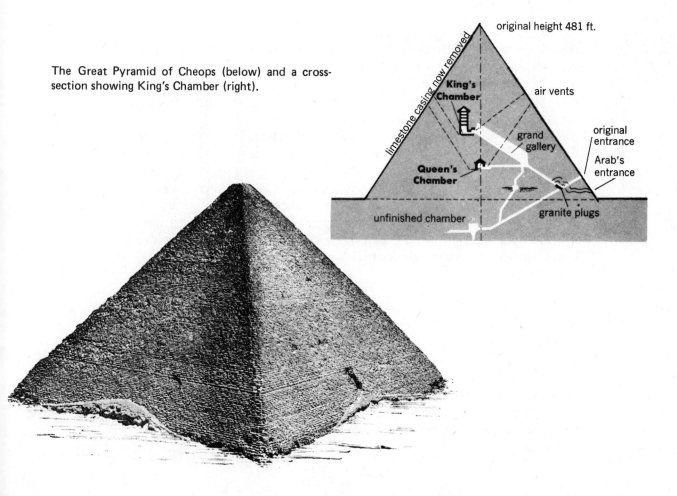

The Great Pyramid of Cheops (below) and a cross-section showing King's Chamber (right).

Recent research by pyramidologists indicates that the coffer in the King's chamber of the Great Pyramid was not used as a burial vault. Rather, it was used by the Egyptian priests as an initiation chamber to test the spiritual condition of disciples and to perform certain feats which would be described as miracles today. A device has been invented by Christopher Hills that recreates the conditions of the King's chamber, enabling us to use the powerful Pi-ray for a host of purposes. It is called the "Pi-ray Orgone Energy Accumulator Coffer." (9) It is designed in such a way that it can only be used for good. If anyone tries to use it for a harmful or destructive purpose that person will immediately experience a return shock. The coffer is described in more detail in Chapter Eight.

Just as there are scientific explanations for the way electromagnetic waves work in our body, there is emerging an explanation of how thought waves work in our mind to link us with one another through the universal field of consciousness and even form the "matter" of this body we wear.

HOLOGRAPHY

Holography is an amazing photographic process which takes pictures in three dimensions. However, these pictures are not optical images, but an interference pattern formed by two or more intersecting wave fields of light. The waves normally used are the two components of a split laser beam of coherent (concentrated in one frequency) light. The first (reference) beam remains unchanged as it is reflected from a mirror to the photosensitive emulsion. The second (subject) beam is changed as it flows around and gathers an informational pattern from the object being holographed. That is, each point on the surface of the object reflects light waves in constantly expanding concentric circles in much the same way that rings are formed when a pebble is dropped in a pool of water. The two waves constructively interfere (without passing through lenses which would form an image) to produce a complex pattern of points surrounded by concentric circles (holes inside holes) on film, hence the name hologram.

After this pattern of intersecting light waves is "frozen" or stored in a photosensitive medium, a three-dimensional image of the object which exhibits all the visual properties of the original object can be reconstructed with light beams. If the original beam of light is passed through the film, the other wave fields are recreated by the effects of the recorded energy pattern. The phase relationships of this pattern create a virtual optical image of the object in space. To an observer this image has the visual appearance of a real object. As participators, however, we know that it is just an image because we can pass a hand right through it. An amazing property of the hologram is that the informational pattern is spread equally over the entire surface. Consequently, any part of the hologram can reproduce the entire image (though with reduced clarity). In the hologram, then every part contains the whole.*

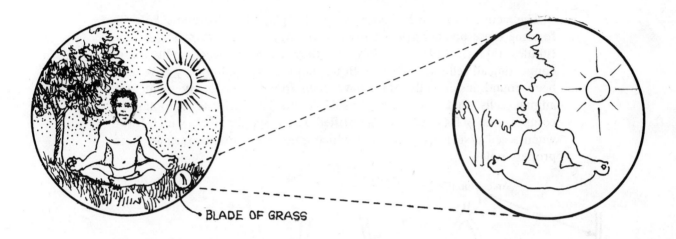

The hologram concept is fast becoming a key theoretical basis for explaining the structure of the universe from micro to macro and for explaining how psi communication works.

THE HOLOGRAPHIC BRAIN MODEL

Several recent discoveries by neurophysiologists, neuropsychologists, neurobiologists, and holographers indicate that our brain works holographically. Nerve impulses travel as waves of energy through as many as 100,000 neurons per second in three-dimensional nerve networks. These networks are capable of storing 10 billion bits of information per cubic centimeter. (28) The only known process which can pack information that tightly is holography. Removals of large amounts of brain tissue interfere very little with perceptual recognition processes and memory. This indicates that perception and memory are functions of the entire brain as if an overall wave field, or mentally-imaged hologram is present. It has even been suggested that echo-locating animals such as bats, porpoises, whales and certain birds use ultrasonic (sound) holography to perform their fine maneuvers and complex spatial differentiations. According to brain researcher Dr. Karl H. Pribram,

---

* As we will see in Chapter Ten this principle underlies the use of witnesses in radiesthesia and supersensonics.

"for the problems of perception, especially those of image formation and the fantastic capacity of recognition memory, holographic description has no peer." (28)

THE HOLOGRAPHIC BODY MODEL

The acupuncture meridians appear to play a holographic role in forming these bodies we wear. In microbiology they are known as microtubules (MT's). (29) Not only do they occur between cells interconnecting all cell nuclei (if Kim Bong Han is correct), but they have also been found inside cells. MT's have been found in so many plant and animal cells that they are considered to be universal cytoplasmic components. MT's have been identified as the spindle fibers which form when a cell divides (mitosis) and which guide the chromosomes to their proper poles.

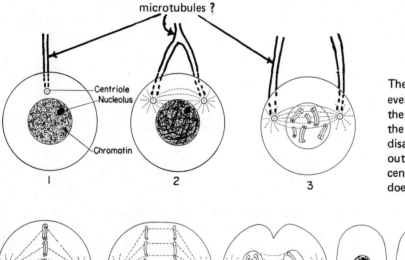

These poles (called mitotic centers) eventually become the centrioles of the two daughter cells. Perhaps at the same time the nuclear membrane disappears inside the cell, the MT outside the cell forms these two centers by splitting before the cell does.

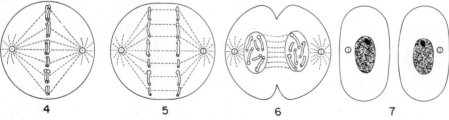

They orient, shape and direct the movement of cytoplasm. They determine cell polarity and are partially responsible for the formation of platelets in blood clots. They build the cytoskeleton around which each cell takes shape (just as they serve as an energy matrix or cosmic blueprint for the formation of our entire body). They do this by transmitting light waves.

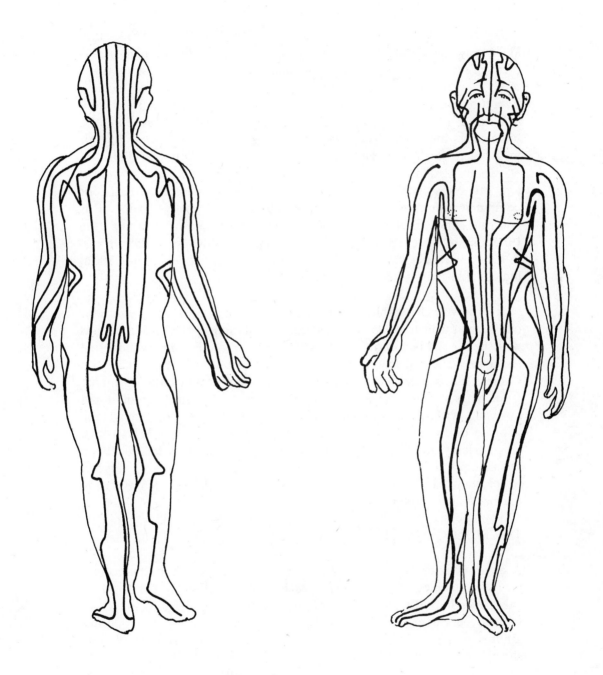

The fourteen most important acupuncture meridians.

Dr. Stuart R. Hameroff of the Tucson Medical Center thinks that MT's may act like fiber optic waveguides for the flow of light waves through our body. (29) He uses the analogy of radar and klystron tubes used in electronics. This theory is similar to the theory of Nuclear Evolution originally published in 1968 which postulated the mitochondria acted in the same way. Both MT's and mitochondria act as geometric waveguides in which electromagnetic waves resonate and dwell, like air in a tuned organ pipe. Dr. Hameroff notes that microtubular dielectric waveguide structures have been well studied in the antennae of moths and other insects. He feels that this provides a theoretical basis for communication among them, especially between mating pairs. This confirms the findings of supersensonic operators concerning the functions of "excited antennae" of insects and dowsers' rods as transmitting and receiving points for subtle forms of communication, including telepathy (see *Supersensonics*).

It is now well established that light enters our eyes and penetrates our brain through the ocular system. Stimulation of the hypothalamus with light via fiber optic implants is found to result in neuroendocrine responses. These discoveries lead Hameroff to suggest that "Chi, the Chinese life energy, is in fact interfering, coherent photic (light) energy from the sun and stars which is refracted by the stratum corneum and resonant within microtubules." He notes that "we now know the neurons of the brain and nervous system are packed with MT's which are at least the right size, shape and configuration to be tuned resonators." In other words, light waves coursing through us are trapped or crystallized by us into energy patterns which act as organizing matrices for our bodily form.

THE BIOLOGOGRAM

Soviet scientist Victor Inyushin presented a paper to the Symposium on Psychotronics held in Prague, Czechoslovakia in September, 1970. In this paper he outlined his idea of the biologogram. Inyushin thinks, as we learned earlier, that the bioenergetic field revealed by electrophotography is a bioluminescent effect of bioplasma. This is a series of plasmic sheaths which interpenetrate our body. It acts as a self-ordering, self-organizing system for the absorption and radiation of energy which remains relatively stable in different environmental conditions. According to Inyushin:

> All kinds of oscillations (vibrations) of bioplasma put together create the biological field of the organism. In the complex organism and its cerebral structures a complicated wave structure -- a biologogram -- is being created, characterized by its great stability as far as the maintenance of the wave characteristics is concerned. (16).

Christopher Hills, in his new edition of *Nuclear Evolution* (1977) refers to this biological hologram as the Universal Hologram.

Inyushin has applied the holographic reality of interpenetrating wave fields of light to the formation of our body. He implies that our body is a three-dimensional *virtual image* formed by wave interference patterns in our mind – our *self-image*. The microtubules may act like fiber optic waveguides for the transmission of mental images in the form of light waves through us. If so, it is quite likely that our body is a creation of holographic thought processes – a *psychic hologram* – *and we self-organize based on the image we have of ourself in our mind.*

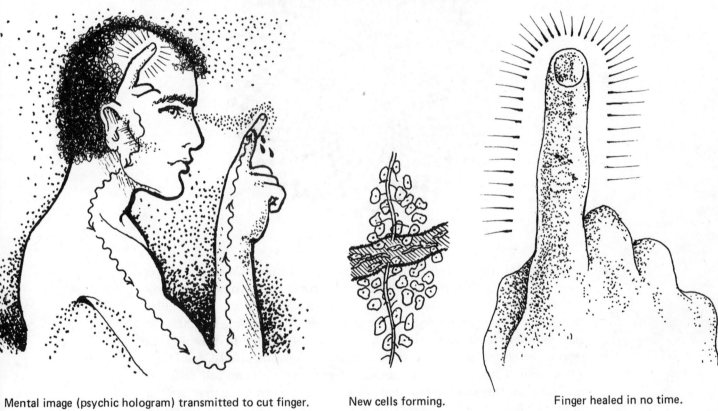

Mental image (psychic hologram) transmitted to cut finger.　　　New cells forming.　　　Finger healed in no time.

This usually unconscious self-image may be transmitted from the brain in the form of light waves flowing to the nucleus of every cell through the microtubules acting like fiber optic waveguides. These lightwaves entering through the cell nucleus would constructively interfere with light waves inside each cell to re-form the image and record it in the genes. This idea springs from experimental work now being done with plasma crystals (see Chapter Six). If the crystalline structure of the gene is in the form of either a circular disc or a multifaceted globe like a cut diamond, a hologram can be recorded in each tiny section or facet. These facets are probably arranged in a spiral formation, following the principle of spiral order found in nature, such as the DNA molecule. Represented visually, this formation would correspond with the symbol of Nuclear Evolution as designed by Christopher Hills.

In addition (as we learned in Chapter One) during mitosis the cell is found to glow with light, becoming luminescent. The chromosomes may be electrified by the action of the mitochondria which act like cellular transistors. When certain crystals are electrified they become luminescent and phosphor reflective. Perhaps the genes function in this way to record the entire life history of the organism holographically, each time a cell splits. Estimates of the number of genes in a human cell range all the way up to several hundred thousand. If each gene has several thousand facets on its surface, then there are theoretically millions or even billions of surfaces capable of recording miniature holograms. (Holography is the best known method for storing the most information in the smallest space.) This raises the possibility that information from *all* the cells in our body is recorded in the genes of *every single* cell in our body. This would explain the principle of cloning by which a physically exact replica or twin of a body can be re-created from a single cell of that body. In human reproduction it would help to explain how one egg and one sperm combine to form a new human being who is connected by an umbilical cord with the mother just as a cell is connected by the microtubule to the rest of the body. This idea accords with the fact that a tiny portion of a hologram can reproduce the entire image of that hologram and follows the Hermetic axiom "As above, so below."*

The holographic model does not end here. Dr. Charles Muses notes that "all the objects we can observe are three-dimensional images formed of standing and moving waves by electromagnetic and nuclear processes. All the objects of our world are 3 - D images formed thus electromagnetically -- super holograms." (18) Physicist Dr. Jack Sarfatti reasons that "the universe may function as a giant hologram, in which the part contains the whole." (24) How can this be? For the beginning of an explanation we must look at that force which at the level of the atom is the nuclear force, at the level of flowing electrons is called magnetism, at the level of personal interaction is called attraction and at the level of still larger bodies is known as gravity.

GRAVITY AND SPACE-TIME

It is now widely accepted by scientists that gravitational fields "curve" space and "warp" time. The discovery of black holes in space-time by astrophysicists is a quantum leap in scientific conceptioning. This discovery makes Euclidean geometry incomplete as far as a true "universal picture" or scientific world view is concerned. Euclidean geometry is founded upon the basic concept that space is flat and straight. It no longer applies in the curved space-time of Einstein and the new physics, just as we cannot apply the two-dimensionality of a plane to the three-dimensionality of a sphere. Astrophysicist Dr. William J. Kaufman, III, Director of the Griffith Observatory notes that:

> Matter and empty space -- the full and the void -- were the two fundamentally distinct concepts on which the atomism of Democritus and Newton was based. In general relativity, these two concepts can no longer be separated. Whenever there is a massive body, there will also be a gravitational field, and this field will manifest itself as the curvature of space surrounding that body. (12)

Kaufman emphasizes that "we must not think that the field fills the space and 'curves' it. The two cannot be distinguished; the field *is* the curved space!" Not only are spatial relationships altered by gravity, but so is time. Clocks at rest in a strong gravitational field run slow relative to gravitation-free clocks.

---

* The chakras may function like holographic "zone plates." These are converging/diverging lenses used in holography to focus and refract beams of coherent light.

THE BLACK HOLE

Here on earth gravitational effects on our body, our space and our time appear insignificant. In astrophysics, which deals with massive bodies, the curvature of space-time is an important phenomenon. It has led to intensive study of that event which causes the most extreme curvature in space-time yet known -- gravitational collapse of a massive star. There is a stage in the evolution of every star where it collapses due to the mutual gravitational attraction of its particles. This attractive force increases rapidly as the distance between the particles decreases. The collapse accelerates. If the star is massive enough, no known process can prevent the collapse from going on indefinitely.

Since the star becomes more dense and heavy as it collapses, the gravitational field around it steadily increases, and space-time becomes more and more curved into a spiralling vortex.

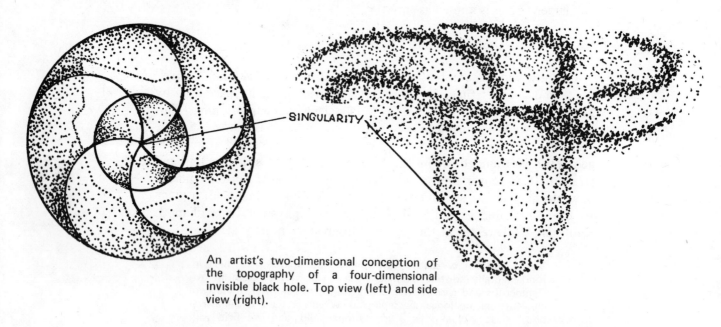

An artist's two-dimensional conception of the topography of a four-dimensional invisible black hole. Top view (left) and side view (right).

Finally the star reaches a stage where nothing can escape from its surface -- not even light! Consequently, we cannot see such a star, because its light can never reach us. For this reason, it is called a black hole. Inside the black hole, some astrophysicists theorize, is a singularity in space-time. Here the curvatures are so large that the local physics becomes drastically altered. At the singularity, *all* laws of physics collapse, and *anything* becomes possible, for singularities are entry and exit points which link different universes together! In the theory of nuclear evolution these singularities are called nodal points. Black holes are regarded in this theory as absolute points where time and motion are stilled and the observer's self-concept is annihilated in the absolute void.

THE SUPERUNIVERSE

Modern physicists now predict that as the black hole grows deeper, the curvature of space-time becomes more and more severe. The eventual fate of this curvature was first described by Einstein and Rosen. They found (theoretically) that space-time opens up into another universe. This "bridge" between our universe and another one was called a wormhole. A two-dimensional representation of this four-dimensional process.

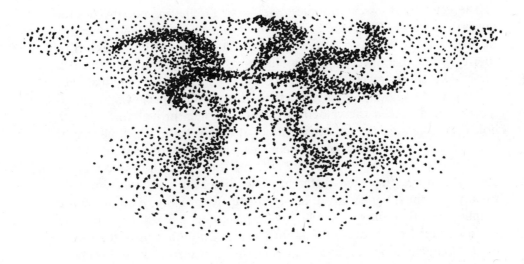

If the upper universe is connected with the lower universe, this wormhole can also be seen connecting different parts of this universe. A very simplified example follows.

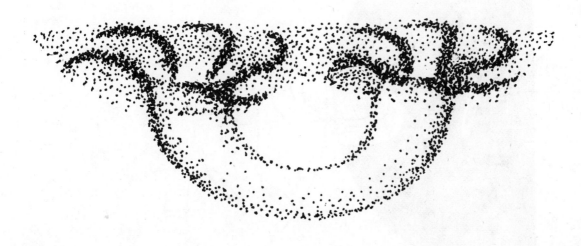

If this huge wormhole connects two points in this universe, what is the result? Many galaxies and quasars emit far more energy than can be accounted for in terms of thermonuclear reactions. Dr. Arp of Cal Tech has discovered a variety of galaxies that appear to be ejecting other galaxies and even quasars. (12) These galactic nuclei appear to be white holes, or the other end of a black hole. In nuclear evolution theory black and white represent the two ends of a recurring spectrum of 80 octaves of matter which disappear into Pure Consciousness and reappear as interpenetrating universes one within the other rather than the wormhole concept.

Theoretically, these wormholes offer some fascinating possibilities for space-time travel. By flying a spaceship into a black hole and emerging through a white hole in this universe we could, conceivably, make trips to its remote parts very quickly. In short, since light travels through these holes space travel is no longer confined to a straight, time-like path limited to the velocity of light in a three-dimensional space. Consequently, we would exceed the velocity of light by going through one of these holes and emerging on the other side of the universe in less time than it takes to snap our fingers. In Einstein's general theory of relativity black holes can act as *time machines*. By entering these black holes we may be able, therefore, to enter different time dimensions of this universe. Finally, and most significant of all, we might fly into any of an indefinite number of other universes which interpenetrate one another and are interconnected inside every black hole. Just as Columbus proved that the earth is curved, not flat, and discovered a "new world" in the process, some astrophysicists now say that space is curved, not flat, and we have found a new *superuniverse* in the process!

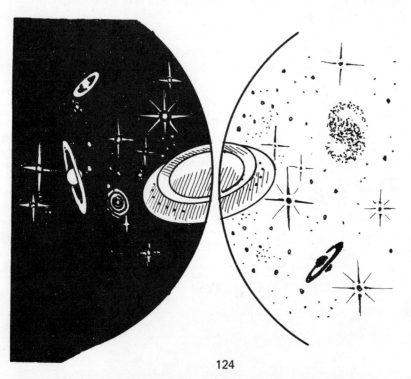

SUPERSPACE

Physicist John A. Wheeler, one of the major creators of the new physics, has followed the hermetic axiom: "As above, so below." He has expanded the astrophysical concept of wormholes in space to a new conception of space itself. Rather than being empty, space is filled -- *with holes!* Wheeler's main image for a new geometry of space is an ocean of foam, where continual microscopic changes occur as new bubbles appear and old ones disappear. *This process is four-dimensional.* Consequently, events from the past or future can exist concurrently with the present. Events in other universes can continue concurrently with events in this one. There is a mix across space, across time, through the wormholes. In the nuclear evolution theory of Christopher Hills, this can already be achieved by entering the holes in our consciousness since all space, time and motion are creations of consciousness.

Physicists Jack Sarfatti and Fred Wolf have elaborated on the imagery of Wheeler to explain the appearance and disappearance of subatomic particles. (24) They think that the bubbles that continually appear and disappear in the geometry of space are mini-white holes and mini-black holes. These tiny holes are *virtual* atomic particles. That is, they only have an apparent, transient existence, being continually created and destroyed. This imagery seems to agree with the dematerialization and etherialization of matter being seen by subatomic physicists. In addition, the strongly interacting particles are reported to rotate in proportion to the square of their mass. (16) This is what is expected if the "elementary particles" are really tiny rotating holes. Finally, since these holes expand and contract around gravitation centers, they refract light waves holographically to form the virtual three-dimensional images we see inside our heads in our consciousness. This imagery corresponds with the discovery by radionic radiesthesiests of the nodal points or the resonance bonds which connect us with the creative forcefield of the cosmos.

These holes, which come in all sizes from micro through macro, explain how everything can be connected with everything else. Sarfatti explains:

> Each part of three-dimensional space is connected with every other part through basic units of interconnection called wormholes. Signals move through the constantly appearing and disappearing (virtual) wormhole connections, providing instant communication through all parts of space. These signals can be likened to pulses of nerve cells of a great cosmic brain that permeates all parts of space. (24)

In this world view, then, superspace is composed of an indefinite number of interpenetrating virtual (apparently and transiently real) universes which interconnect with one another through holes. At the center of each hole -- the singularity -- *every hole is connected with every other hole.* In this new scientific world view, as in the ancient mystical one, *the center is everything.*

CONSCIOUSNESS AND REALITY

Consciousness is the concept which unites mind with "matter" in the new physics of psychophysics and supersensonics. It discards the Ideal I in favor of the real I and returns the Self to its rightful place in science -- the center.

> Consciousness is the totality beyond space-time -- what may in essence be the real "I". We have come to know that consciousness and energy are one; that all of space-time is constructed by consciousness; that our normal perception of reality is a composite of an indefinite number of universes in which we coexist; and that what we perceive as ourselves is only the localized projection of the totality of our true selves. (24)

In psychophysics "matter" is considered as a construct of thought taking place within the total field of consciousness: "vibrations of thought patterns in specific harmonies structure all 'matter' and light as we experience it." Every thought we have, therefore, influences everything that happens in all the universes. We are responsible for our thoughts. In supersensonics even thought is considered to be a form of subtle matter since mind energy is never free of the thinker of the thought.

The here and now is linked with the beyond through singularities or absolute points of timelessness. As expressed by Sarfatti: "singularities are entry and exit points of that which is beyond space-time projecting itself into space-time. Space-time self-destructs at these locations (points) in the gravitational field." Bob Toben (creator, with Sarfatti and Wolf, of *Space-Time and Beyond*) notes that:

> Beyond space-time is nonphysical, unmeasureable. At the singularity all laws of physics collapse. But what is beyond space-time is within everything. Can it connect with us and influence us within space-time? Is it PURE CONSCIOUSNESS?

Sarfatti conjectures that interpenetrating our body is a self-organizing, biogravitational field which *generates* the body from that universal field of the light of consciousness. He reasons that "a participator in a high state of consciousness can," he feels, "create black holes and white holes in his local biogravitational field." Therefore, "consciousness may be able to alter the patterns of constructive interference to create separate but equally real realities." This means, in essence, that we can experience other worlds if we wish, by purifying and expanding our consciousness. This corresponds with the Yoga teachings concerning the opening of the crown chakra -- the way to liberation in union with the ALL. The existence of other worlds or realms of existence called lokas is a major tenet of Yoga and other mystical teachings. The way to the Whole is through the hole at the center of all being. Union with the ALL is now a scientific as well as a mystical teaching!

DIRECT PERCEPTION

The faculty of direct perception is the means by which, through the eye of the ONE, the true nature of existence and nonexistence is seen and known. But the ability to see through this eye only manifests when we go beyond seeing with the Ideal Eye of the Ideal I. Einstein created a *theoretical mathematical model,* an "ideal observer" who sees the universe the same for *all* observers, regardless of position (whether at the center or periphery of a system). To Einstein, then, there were no favored positions nor any boundaries. This perspective creates a *centerless* system. Although it frees us intellectually from the perspective of an ego occupying a body, it is incomplete. It places us outside reality, rather than in it. From experience we know that the view from the center of a system (our body for example) is quite different from the view at the periphery (for instance a television picture of our body). By standing at the center of the system observed, we create a universal observer *both within and without,* freed from the constraints of time and space. This is the link between the seventh level of consciousness -- the creative imagination -- and the beyond, the prime imagination of the guiding logos or divine plan. By and through this link at the center of our being, visions become realities. They filter down through the various levels to the physical as we open our imagination to them. The eye of the ONE sees ALL in and through ALL. It is the centerless center which is everywhere, everywhen. According to Christopher Hills:

> Relative to the center there is no motion in reality because in nature everything is actually moving with the center. But relative to an artificial theoretical observer situated outside the total universe being investigated, the velocity and motion is created by our consciousness of being separate. The illusion is carried forward into the experience of objects, because again and again we think ourselves separate from the universe we live in and the objects inhabiting it. Only when we can permanently come to see that consciousness extends around all these objects, which we call objective, and see that they are the center, in other words, really experienced inside consciousness and not external to it, can we experience the flash of illumination. (10)

AT THE ZERO POINT

Cosmologist Arthur Young has recognized the fundamental importance of consciousness as a basic ingredient of the universe which must be included in any macrotheory about it. He asserts that:

> Because we have a knowledge of consciousness at least as certain as our knowledge of time or space or mass, it must be given recognition. Since no scientific observation can be made without consciousness, the latter cannot be accounted for as a compound of lesser ingredients; since it is thus basic, our cosmology must be revised to include it. (18)

And this is just what he has done.

The most elegant and simple formula for the volume of the universe (called the Einstein-Eddington hypersphere) is $2\pi^2 Y^3$. Young notes that this formula fits the shape of a torus (donut) with an infinitely small hole at the center. He lists a number of reasons for a toroidal-shaped universe: "the magnetic field, the vortex, and the tornado all have a toroidal form, and the vortex is the only way a fluid can move on itself." This is the same way that many mystical teachings describe the universe -- as a giant mandala or spiralling wheel, containing a myriad of smaller wheels. In nature we see all around us the truth of this idea. We see it in water as it swirls down the drain and in the ocean we call it a whirlpool. We see it in fire as it follows the updraft in a spiralling spin. We see it in the air in the guise of tornadoes, typhoons, cyclones, and hurricanes.

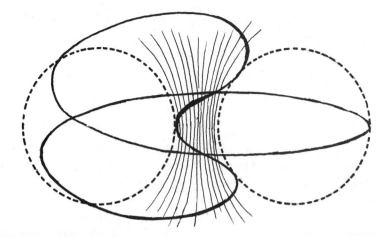

The motions of a vortex (left), and as reflected in Nature (below).

We see it too in the mushroom spiral of the nuclear explosion. It is recorded in "matter" in the growth processes of living organisms, for example the spirals in shells, and the growth rings of trees. Everywhere we look it seems, we see circles inside circles, spirals inside spirals, spheres inside spheres, tubes inside tubes, holes inside holes. Our body itself is full of holes. The skin is penetrated by unknown billions of holes of sweat glands and hair follicles. Tunneling tubes of arteries, veins, capillaries, nerves, meridians, microtubules, etc., crisscross our body in intricate patterns linking every cell with every other cell. We are a food tube (the digestive system), a *continuous hole* into which passes nourishment and out of which pass wastes and toxins. The chakra system, too, can be viewed as an interlocking system of holes sucking in nourishment from the universal field of the light of consciousness. When all the chakras are balanced in stillness they create a "tunnel" for the passage of kundalini (consciousness), the divine nectar of transcendent bliss. Each cell is a hole or sack-like receptacle dependent for its continued existence on something larger that supplies it with nutrients, informational messages, psychic electricity, etc. Likewise, we too owe our existence to something larger that nourishes us. We cannot really separate ourself, at any level, from our larger Self, our real I.

The toroidal form solves the philosophical problem of the individual versus the cosmos, the part and the whole. The part *is* the whole. We only delude ourself by *thinking* we are a separate being occupying a body. In a toroidal universe any part seemingly can be separate and yet be connected with the ALL through the infinitely small hole at the center, which is within everything, everywhere.

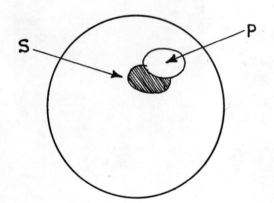
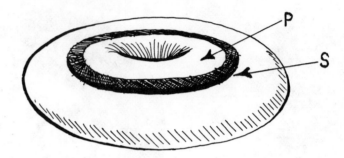

In a sphere (left) a part (P) can be separated (S) from the whole. In a torus (above), however, a part can be seemingly separate but still connected with the whole through the infinitely small hole at the center.

The centerless center has again been glimpsed! The Cosmic Hole leading to the Cosmic Whole is now a central scientific teaching as well as a mystical one. Science and spirituality are joined, the experimenting scientist and experiencing mystic become one in us as we jump into this hole -- the center of centers, the Cosmic Zero.

> Zero was born an absolute. Zero consciousness of self was the essence of the Universal Self who was present in all seeds of personal selves. The law of its being was that all radiating matter which we experience as light, seeks continuously to annihilate itself in the lower pressure zones of gravitational centers. The light of creation was always seeking holes. The creative force which activates the creation always ends in the empty womb of space to be continuously born again as the Self -- the One who is transcendent and immanent, all-present in every wave of light and vibration of sound. The cosmic song of Self is simultaneously born in every desire or attraction and dies in its fulfillment of the orgasm of self-annihilation. For some, the path to this knowledge of Self is curved in direction but spiral in motion: for those avatars who know this Truth directly, it is straight penetration into the void because in them there is no direction and no motion. There is only the ONE who sees all through them.
>
> --Christopher Hills
> "Supersensonics"

EXPERIMENT SECTION FOR CHAPTER FIVE

A PSYCHOTRONIC ENERGY DETECTOR

After practicing the experiential techniques outlined in the previous chapters, you may have begun to feel psychic electricity, psychotronic/bioplasmic energy, Qi, prana or whatever you wish to call it flowing through your body. Now it is time to provide visual proof that this energy has physical effects. The ability of our mind to influence matter directly is called psychokinesis. We can demonstrate this phenomenon by constructing a very simple device. All you will need for this experiment is one hollow, wide mouthed glass jar, one piece of very thin thread (preferably silk), one two or three inch long pig's bristles or straight, stiff hair, one toothpick and two small dabs of fast setting glue.

The construction procedure is as follows: cut the thread to a length half as long as the jar is high. Dip one end in a bit of glue and place it against the bottom middle inside the jar and hold it there until the glue sets. Invert the jar. Now dab a bit of glue on the dangling end of the thread with a toothpick. Attach the bristle. Now place the jar with open mouth against a flat surface to insure that no air currents can influence the experiment. Also keep it away from heat sources so that convection currents do not interfere.

STEP ONE: Sit in a comfortable position with back erect and spine straight. Breathe slowly and deeply to relax and charge yourself with energy. Now focus your eyes and your concentrated attention on the bristle and visualize it in your mind as beginning to turn slowly in a clockwise direction.

STEP TWO: While continuing to look intently at the bristle with a vivid mental image of it turning, place your hands on opposite sides of the jar at a distance of about one or two inches. Visualize a stream of energy flowing from the palms of your hands and forming a vortex or whirlpool which turns the bristle in the clockwise direction you have visualized in your mind. It may help to move your hands around the jar in a clockwise direction. Do not force the action. Just remove any obstacles in your mind which prevent the image from forming or the energy from flowing. It may require repeated practice for a couple of minutes or it may happen for you immediately. Concentrate.

STEP THREE: This is a more advanced technique to practice after you achieve success in step two. Instead of using energy from your hands, this time use the energy streaming from your eyes, the eyebeam. Again form the mental image of the bristle turning in a clockwise direction. With total one-pointed concentration send that image out from your eyes on the eyebeam. It may be helpful to practice alternate nostril breathing and mirror gazing to amplify the eyebeam.

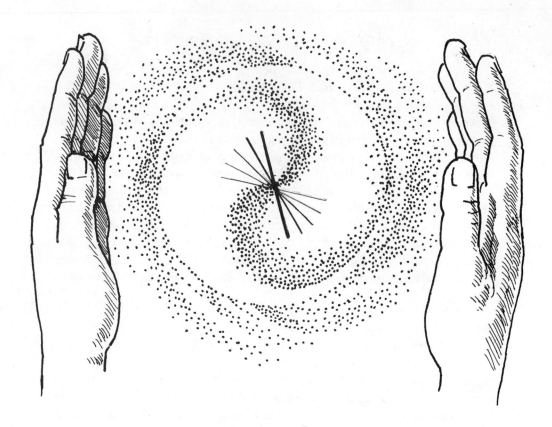

This device for detecting psychic electricity is paradoxically similar to the first device for measuring the flow of electric current invented by Volta. He used a pig's bristle inside a sealed tube to measure the flow of electric current and voltage was named after him as a result.

You can use your imagination to devise similar experiments. But remember to eliminate the influence of inertia and friction as much as possible. The less energy needed to overcome resistance the easier it will be. The movement or rotation of a cork in a large container of water is relatively easy. A large container of water such as a bathtub is preferred in order to reduce the effects of surface tension inherent in liquids. Or you might try sticking a needle into a small paper straw and embedding the other end of the needle in the cork. Try rotating the straw in either direction with the energy flowing from your palms while holding a distinct mental image of it turning in the desired direction.

TELEPATHY EXPERIENCES

Scientists at the Toronto Society for Psychical Research have found that rhythmically pulsed thought waves aimed at a participator isolated some distance away "evoked responses in the EEG which are similar in form and comparable in magnitude to those evoked by physical stimuli such as sound." (19) In other words sending of a telepathic image in waves has been directly connected with changes in the brain wave patterns of the receiver. This corresponds with the findings of Soviet scientists that telepathic messages are best sent in rhythmic pulses of thought waves tuned to the heartbeat through breathing exercises.

You can construct a simple device for sending messages telepathically which uses a light that flashes rhythmically. All you need is a light fixture with a 15 to 20 watt bulb and a 1" in diameter by 1/8" thick flasher button which is available at most hardware stores. Merely unscrew the bulb and insert the flasher button into the light socket after first disconnecting the lamp. Reinsert the bulb and plug in the lamp. When the lamp is switched on the flasher will turn the light on and off at the rate of about 70 times a minute. This is very close to the rate of the average heartbeat.

If you prefer a slower, more relaxed rate there are also winker plugs available at hardware stores which go between the plug and the wall socket. You can also make a portable device by using a battery operated flasher lantern, a flashlight with a flasher switch or an auto safety flasher unit available at many automotive parts stores. In any case be sure to use a light that is easy on your eyes, and remove the red cover if it is an emergency beacon.

Next you will need to prepare a series of opaque cards or thin slips of paper upon which to place the images you will be sending. For images begin with something simple such as numbers, letters of the alphabet or simple geometrical designs such as circles, squares, triangles etc. You might try colors as well since colors have different feelings to them and some people find them easier to send and receive. Colored geometrical figures such as a green triangle, a red square, a white circle etc. are recommended for they combine the feelings of shapes with the feelings of colors and hence amplify each other.

After placing on the cards the images you wish to practice sending, fix a stand of some kind above the lamp (if looking down from the top) or in front of it (if the lamp is at eye level). This stand should be designed to hold the cards between your eyes and the flashing light.

To make the preparations complete, you will need one or more of your friends to act as receivers. The more people that participate the more fun it will be, and be sure to take turns sending and receiving. Provide the participants with sheets of paper to write down the images they receive in the order that they are sent, or take turns as scorekeeper for the entire group. At the beginning be sure to let everyone know what set of images is to be sent. Later, after some practice, you may wish to make the task more difficult by keeping them secret. Also be sure that the receivers are somehow screened from the pulsating light, especially if you are sending colors. Instructions for the sender follow.

FLASHER BUTTON

This device can also be constructed using a black light bulb. However, instead of shining through the card, the light must be reflected onto it from a mirror-like surface. Use day-glo or special fluorescent paints to make the symbols and they will glow brightly in the dark when irradiated with black light.

STEP ONE: Sit erect in a comfortable position, if possible facing in the direction of your receiver(s). Close your eyes and begin breathing rapidly through the right nostril. Energize yourself mentally and physically from the tips of your toes to the top of your head. Focus your concentrated attention on the ajna center in the middle of your forehead where telepathic images are both sent and received.

STEP TWO: Erase all doubt from your mind, open your eyes, turn on the lamp and focus your concentration one-pointedly on the image flashing in front of your eyes. Chant the name of the image silently to yourself in rhythm with the light flash. If the image is a color, violet for example, chant the word violet rhythmically with the flashing light. Surround yourself and immerse yourself in violet. Visualize nothing but violet on the screen inside your ajna center. Now release that color through the black hole at the center of your being. Send it consciously in a precise direction and to a definite location -- the mind of your receiver(s). Remember if you as sender do not form an image or thought clearly the receiver will have trouble picking it up. Concentrate.

The receivers should practice the following at the same time as the sender is doing the above.

STEP ONE: Sit or lie down in a comfortable position. Close your eyes and breathe deeply and slowly through the left nostril. Relax mentally and physically from head to toe, but remain alert with mind cleared of extraneous thoughts.

STEP TWO: Erase all doubt from your mind. Visualize a blank screen in your head in the indigo ajna center in the middle of your forehead. Wait patiently for an image or intuitive sound to form in your mind from your center. Try not to guess. Note the first image or sound that pops into your mind. Write it down or tell the scorekeeper.

Just as several batteries connected together produce more power than they do singly, so can telepathy be enhanced in a group. A sort of psychic power circuit, or group mind is formed when all minds in a group move in harmony with each other. To test this out with your group, merely set the flasher unit in the middle of the room and have the entire group rhythmically send the images to a single receiver in another room. If you decide to send a picture, make it a striking scene with vivid details and a strong emotional theme. Emotion adds power to telepathic messages.

Call the receiver into the room and ask him or her to pick the scene from three or four pictures. This is a good way to test your effectiveness.

You might wish to divide your group into pairs of senders and receivers and give them different images to transmit. Do the messages cross or do the receivers get the message intended for them? Who is the most effective pair? Who is best at transmitting complex or simple images?

Experimenting will show that telepathy can be selective. Just like radio sets, people tune in to different channels. When there is only one sender and several receivers telepathic thought pollution can occur. That is, a receiver may be picking up a false image in the mind of another receiver rather than the true image. It may help to ask receivers to surround themselves with circles of white light with the suggestion that only the true message will get through from the center.

In any event group telepathy experiments can be fun if conducted like games in a friendly spirit of cooperation. Practice will indicate those with natural telepathic ability and may inspire everyone in your group to explore their supersensitive nature more fully. *Conduct Your Own Awareness Sessions* (now republished as *Exploring Inner Space*) provides a series of gamelike group activities to enhance sensitivity, expand awareness and increase caring and sharing.

THE MERKHET AND OSIRIS PENDULUMS

The Egyptian word "merkhet" means instrument of knowing. This vertically shaped pendulum, filled with mercury in its hollow cavity, was used by the ancient Egyptians to amplify the detection of the direction of stars, the orientation of temples and pyramids and for surveying the Nile lands after flooding, through mental questioning. They believed that the answers came from the god Thoth, which is the equivalent of the Greek god Hermes or the Roman god Mercury: the principle who governed messages of the mind.

In more scientific terms, mercury is found to resonate with thought waves. The Merkhet and Osiris pendulums can both be used for measuring the I.Q. of a person or the intelligence of animals as well as what they are thinking. They can detect the thought patterns and karmic memories. By using a mercury filled pendulum we can find out if a relationship is beneficial or harmful to us. The Merkhet is made from a special high impact material for rough use. The Osiris is handmade from transparent glass through which the mercury shines with a beautiful silvery glow. For more information on the blending of wave fields read Volume III, *Supersensonics*.

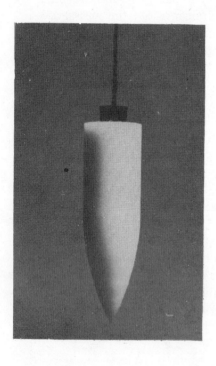

THE MERKHET

THE OSIRIS

SUPERSEEING

THE WORLD CULTURE

Now that we have jumped through the hole and have conceived a new universe in a new space and placed a new mind behind new eyes in a new body with an expanded consciousness, let's supersee our supersphere. We seem to be in a space that is filled, not empty. It is filled with light in the form of swirling, dancing, kaleidoscopic energy patterns of images, archetypes, symbols, effigies, phantoms and mirages. We see a shifting panorama of constantly changing ethereal forms, interspersed multidimensionally in the interpenetrating superconscious of Universal Mind. Images appear from nowhere and disappear to nowhere. Images we can see, but cannot touch. They pass through us and we through them! Macroscopic, microscopic, multiperspective timeless superseeing fills the Cosmic Void with Cosmic Consciousness. We have not lost our mind, we have found it! Are we free, at last?

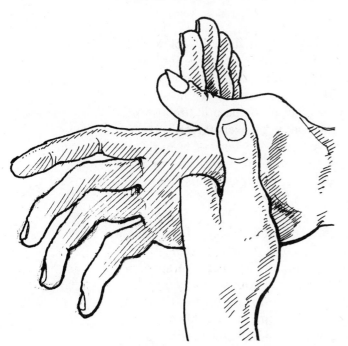

The world of images is the realm of the mystic, the artist, the poet. While the scientist in us stands agape as "matter" dissolves right in front of our eyes, the mystic at our center just smiles. The positive, rational, masculine, solar, Yang, Shiva scientist and negative, intuitive, feminine, lunar, Yin, Shakti mystic (in essence all complementary opposites) are beginning to merge in supersensonics,* the manifesting of our most secret thoughts and wildest dreams right in front of our eyes. If seeing is believing, then many of our most cherished and deeply held beliefs are about to change.

---

* In this context, supersensonics denotes the combination of applied psychology and radionics with technology, psychotronics with electronics -- the technical medium for stimulating mass transcultural awareness of our evolving planetary consciousness.

A powerful combination of inventive audio-video technology and imaginative videocinematic techniques is creating a new art form. This integration of the latest scientific technology with practical mysticism promises to become an all-encompassing world phenomenon -- a world culture -- leading to the teaching, practice and experience of the same spiritual science everywhere by nearly everyone.

Our constantly growing multi-mass-media communications systems are circling the planet, enmeshing more and more of us in a growing web of interconnections. These channels of information-flow serve as nothing less than the nervous system of that evolving planetary being which includes each of us, like cells in a great planetary brain. Television is especially important because it provides for a vast expansion of our visual imagination. It crosses boundaries, interpenetrates local cultural tradition and provides a shared cultural experience for millions of us daily. It expresses and transmits our symbolic needs, enlarging not only our physical world, but our psychic one as well. No longer can we escape the human condition, for we see it daily on TV. Televison shows the human species to itself as a working model of itself. It reveals human consciousness to human consciousness and expands our awareness that we are a planetary being.

PLANETARY BEING

Just as the nucleus in every cell in our body is directly connected through the brain wholistically with every other cell, so are we inseparably interconnected with one another and all life -- though we may not be fully aware of it. To realize our deeper universality we need only surrender our illusory separative self-sense and participate in life instead of watching it pass us by. Through the center of every system -- the nucleus of every atom, the nucleus of every cell, the heart of everything -- everything is connected with everything else. So it is with us. We are cells in a much larger being interconnected through the universal field of consciousness.

The folly of our individually throwing fists and collectively fighting wars is surpassed only by our ignorance of our common origin. If we trace our ancestry back to the beginning of life on this planet, it appears obvious that we all spring from a single, common source. Projecting the evolution of life (consciousness) on this planet into the future, is it not likely that we all return to that source? Looking at our life now, it appears that the only thing that separates us from that source is our idea or mental image of ourself as a separate entity, an isolated being, a detached ego, a relative self. If we hold an image of our highest expression as a relative self and not an Absolute Self, we will never understand how energy manifests through our being from the center, our heart. Energy manifests through our deepest center in direct proportion to our level of consciousness.

THE EVOLUTION OF CONSCIOUSNESS

The blending of inner with outer, real with ideal and true science with true spirituality is becoming more closely achieved as scientist-mystics turn their attention and highly developed tools of mental investigation to the study of consciousness. The late scientist and Jesuit priest Pierre Teilhard de Chardin in his classic work *The Phenomenon of Man*, formulated a cohesive, factually substantiated and very convincing theory of evolution. He started from the position that the human species is a totality, a planetary being, that can be described and analyzed like any other phenomenon. We and all our manifestations, therefore, including human history, human values, and human consciousness are proper objects of scientific study. Like Darwin, he realized the dynamic self-transforming nature of life and the necessity of adopting an evolutionary point of view. He saw the entire universe as a process of becoming, of attaining new levels of existence or higher self-organization through the conditioned genesis of conscious adaptive mutation (willful change). Just as Einstein's vision expanded the classical physics of Newton to bring the new physics of psychophysics, so did de Chardin eclipse the thought of Darwin to bring us psychobiology.

He reasoned that evolutionary fact and basic logic demand that as bodies evolve, so does awareness. Consequently, by following the evolutionary spiral backward from the human to the biological phase of existence and from there to the inorganic phase, we can logically infer the presence of potential consciousness in all material systems. He maintained that as organisms convergently integrate into increasingly complex units of self-organization there is an intensification of mental activity, the evolution of progressively more *conscious* mind.

This reasoning, he felt, could be applied at the species level. Just as unknown millions of cells organize themselves into what we call our body, so are over four billion of us organizing ourselves into a larger body, a planetary being. This process is accompanied by the development of the technological equivalents of sense organs, effecter organs and a coordinating central nervous system with dominant brain conceived and built in the image of our own body. Thus the spread of modern mass communications systems is inexorably sweeping us into conscious realization of our part in this being as it helps us raise our consciousness to the planetary level -- planetary consciousness. This is a worldwide transcultural phenomenon which is leading to the gradual union of the family of humanity into one interthinking, interfeeling entity based upon a single, self-developing framework of thought about consciousness.

Christopher Hills in *Nuclear Evolution* describes eight spheres or layers of consciousness as conceptual tools for understanding that which is a nonconceptual reality. Each sphere represents a level or degree of separateness of our ego-self from the whole. The first three correspond with the id, ego and superego of Freud and the "self" of Jung. They are: body consciousness, ego consciousness and self-expressive consciousness. When these three are well integrated a centered self emerges which is able to progress to higher levels. In order of development they are: acquisitive consciousness, family consciousness, community or group consciousness, world, global or planetary consciousness and cosmic or universal consciousness. Hills notes that:

> When the universal consciousness is reached the personality realises the unity in all things, and becomes conscious of the Spiritual Sun with its multidirectional rays, and all our spheres of individual consciousness are transmuted into radiating spheres of light. The next step after realisation of this is its manifestation in conscious Nuclear Evolution.

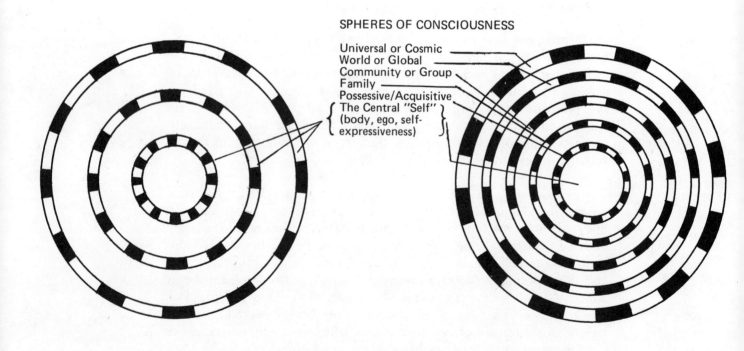

THE PSYCHOSPHERE

Psychosphere is a conceptual term to indicate the vast scope and influence of television on a planetary scale in many simultaneous fields of sense extension. New developments in the supersensonics of expanded seeing are laying the groundwork for an interconnected worldwide communications system that will be out of this world. The psychosphere is telepathy on a planetary scale. Through the universal field of consciousness we are linked together into a coherent, synergistic whole – like cells in a great planetary brain. Energized thought is a powerful force when concentrated and consciously directed. Our multi-mass-media communications systems are stirring millions with new ideas and visual experiences unparalleled in human history. As certain insights into the nature of consciousness capture our collective imagination, we rise together to higher levels of self-organization, eventually spanning the superuniverse with our expanded consciousness.

OUR PLANETARY NERVOUS SYSTEM

There are over a quarter of a billion television sets in use around the Earth. A system of direct, satellite to home TV is on the way. Many events are already being relayed via satellite for home viewing at the same time they are occurring thousands of miles away. Audiences numbering in the tens of millions are viewing history as it occurs. The COMSAT system which is composed of about fifty domestic communications satellites is capable of carrying over one hundred thousand TV channels and over one hundred million voice channels. (27) The scores of satellites in the military surveillance systems are capable of near ground level photos of nearly any exposed place on the entire planet with registration and color qualities as good as National Geographic. These systems give us the ability to see as we have never seen before.

Television and the world multi-mass-media systems interpenetrate local cultural tradition providing commonly shared experiences in unprecedented ways (especially among children). This global network allows us to share and transmit our symbolic needs and their expression on a world scale, creating a world culture. Inventions like television have extended our collective perception until we can now telescope time (for example, time lapse photography), move into the past (for example, videotape recordings) or into the future (for example, artistic simulations) and span the whole Earth at a glance (for example, satellite photos of weather conditions). New inventions are on the way to speed our evolution by extending our perception still further.

COMMUNICATING WITH LIGHT

Laser fiber optics is a burgeoning new communications technology. Light pulsed from tiny lasers at extremely high frequencies is transmitted through a system of fiber waveguides, carrying audio and video information. These fibers are now being used by medical scientists to see inside our body (films of which have been shown on TV). Each tiny fiber cable is capable of carrying thousands (theoretically even millions) of voice and picture channels. Since they are made from sand, they are relatively inexpensive to produce. Most of the costs of worldwide installation will be covered by recycling the now obsolete copper wire systems which presently crisscross the planet.

These fibers have several technical advantages. They are immune from stray electromagnetic fields. They resist interference and "crosstalk" (signal coupling between lines) is cut to zero. They are almost impossible to "tap" and thus provide greater privacy and security.

The lasers used to pulse the light waves through the fibers are quickly being perfected. Scientists at Bell Laboratories have already created the world's smallest lasers. Their size is less than that of a grain of salt. These "injection lasers" can be flashed at the incredible speed of half a billion cycles per second. This makes fiber optics by far the fastest known information transmission system. With this system a twenty-four volume set of encyclopedias can be transmitted, letter by letter, in less than a tenth of a second!

The first fiber optic telephone system is already operating between New York City and Newark, New Jersey. Bell scientist John Douglas calls laser fiber optics "a fundamental change in communications, the most revolutionary since development of the communications satellite." (27) Because of its many advantages, it is fast becoming the basis for a greatly expanded planetary awareness. Thanks to another recent invention this system will soon be used to transmit visual images in three dimensions!

THE PLASMA CRYSTAL SCREEN

This invention arose from the discovery that "certain liquid crystals can be made opalescent, and hence reflecting, by the application of electric current."* (27) These plasma crystals, when specially arrayed in the form of an electrified screen and fed by beams of light via fiber optics, can reproduce visual images with real depth, without the need for 3-D glasses. These screens can be made any size, from wristwatch to billboard! In addition, the clarity is reported to be bright enough to be visible in direct sunlight. According to Youngblood, these screens are scheduled to go on sale commercially in 1978. (27)

The implications of such a visual "hole in space-time" are truly vast. Imagine a wall size window on the world linked directly via satellite with all the radio and TV stations on the planet. Add electron microscopes, telescopes, sonarscopes, TV cameras in orbiting satellites and on planetary probes, etc. All these electronic and technical devices together represent a quantum leap in the expansion of our seeing. They are bound to have deep psychological effects on the captivated millions awaiting

---

* All the cellular fluids in our body have plasma crystal components.

the new visions which are about to unfold right in front of our disbelieving eyes. Some of these visions will come with the aid of that amazing reservoir of information and replica of the human brain -- the computer. But, however sophisticated the means of communication is, it is still the message and not the medium which actually changes us. To confuse the medium with the message is to see only surface changes in our life. Without changes in consciousness there is no creative advance of intelligence.

THE COMPUTER WORLD BRAIN

Some scientist-artists are turning to the computer as a creative medium which has already reached a high degree of technical proficiency, dependability and versatility. Computer capabilities are expanding rapidly. New teaching machines approaching speeds *one million times faster* than the fastest digital computers are on the way. (27) Each year the number of computers in the world is doubling. At the same time, thanks to microminiaturization, computer capabilities are increasing by a factor of ten every two or three years. The next few generations of learning machines will be able to perform in five minutes the work that would take a digital computer ten years to complete. Computers are being joined together in networks for rapid access to the latest scientific and mystical information (computers are now programmed to do astrological charts, for example). The computer's great speed, freedom from human error and vast ability for producing the unexpected (when set in random mode) are great assets for artists when the computer is given a video imaging ability.

Synergetic synthesis of computer-video operations is well underway. Computers are being programmed to simulate cameras and the effects of other conventional filmmaking procedures. Computer generated images in motion already appear on many television stations regularly. Soon 3-D figures will be appearing on a 3-D screen, when the computer is successfully linked via fiber optics with the plasma crystal screen. Artist Eric Siegel is aware of these developments and comments that:

> I see television as bringing psychology into the cybernetic twenty-first century. I see television as a psychic healing mechanism creating mass cosmic consciousness, awakening higher levels of the mind, bringing awareness of the soul. (27)*

---

* The Electronic Soul Mirror is a psychovideo technique for a greater self-awareness used at the University of the Trees. It consists of a videotape of oneself in creative conflict with others. As one watches the playback, one is again videotaped. Then one watches oneself watching and sees oneself in a new light. We transcend ourself and perceive ourself in the act of perception.

How can this become possible? It will only happen when we *directly* tap into the creative imagination and through it to the Prime Imagination, the guiding logos or divine plan which silently and invisibly guides the evolution of our consciousness.

THE HOLOGRAPHIC IMAGINATION

The final link in this mind-blowing chain of supersensonic innovations is holography. As we learned earlier, a holographic paradigm (world view) is emerging in science. It is almost a living reality in the art of videocinematics. Still pictures (holograms) are already selling commercially (as covers for greeting cards, for example). The first holographic movie has been made. Although still in its infancy, when perfected holography will become a fundamental reality of our superseeing. Youngblood notes that:

> The ability of holography to record natural phenomena that exist beyond the range of human perception -- shock waves, electrical vibrations, ultraslow motion events -- could contribute to an experience of nonordinary reality beyond the reach of conventional cinema and television. Like Alice in Wonderland, no longer looking at a vision in a mirror but being in it. (27)

Holographic TV is on the way. According to the December 1976 issue of the "Videocassette and CATV Newsletter," the Japanese television manufacturer Hitachi has announced the development of a low-cost holographic videodisc system.

> Fifty-four thousand very small holograms, each one millimeter (0.04 inches) in diameter, are spirally arrayed on a 30 centimeter (11.8 inches) transparent vinyl chloride disc. A laser beam is aimed at the hologram to read out the stored information. One hologram is capable of storing information which corresponds to one frame of a television picture including sound, and each disc has a 30-minute playing time. (33)

Imagine a projection system that does not use a screen to form images. Instead, a beam of light records an image from a holographic TV camera and passes through an optical fiber to a home projector. This projector, mounted on the floor of your living room, for example, *re-forms* the image in a full three dimensions in the space directly above it.* Now if this image is that of a person, for all visual intent and purpose, that person will appear right in front of your eyes. The only difference from "normal" will be that if you try to touch that person your hand will pass right through the image as if no one were there, although you will be

---

* This would be the equivalent of the Star Trek "transporter" right in our own homes (though it would send an image of our body instead of turning our body into light and sending it.).

able to see and talk to the image. If at the same time that an image of a friend of yours is projected into your home an image of your body is projected into the home of your friend, both of you will be seeing each other and conversing together *in both places.*

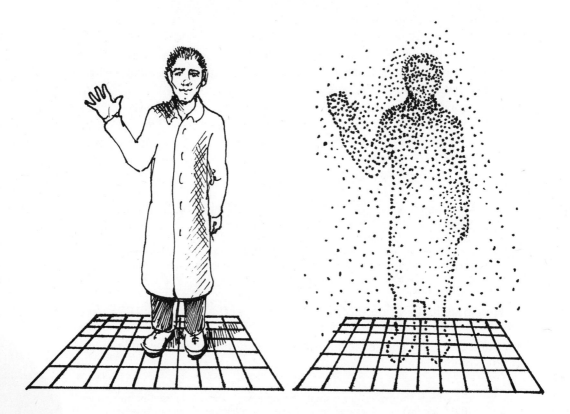

This visual effect could lead to a change in consciousness. If, as a result, we become less attached to our notion of ourself as a being occupying a body or a specific spatial location, then we will become more aware of that universal field of consciousness which unites us.

Stretching our imagination a lot further into the realm of pure speculation, maybe we can envision a supersensonic invention which will make the next logical interface between holography and the imagination. Optical fibers are now used to peer into our body and take pictures of our "insides" with light. If this process can be reversed and a fiber is inserted into the nervous system or acupuncture system, maybe we will be able to observe the light waves passing between the brain and the cells. If these light waves can be shunted into a holographic video display system through some future psychotechnical ability, then everything we can visualize in our imagination could appear right in front of everyone's eyes!

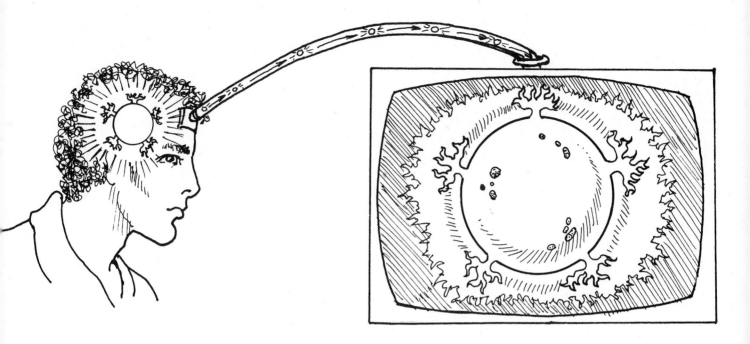

However, this in itself is only an effect of a mechanical video process which will not make us more original or creative, but only more prolific. We must go beyond the sensory illusions, the dance of Maya, to the source of vision -- consciousness itself. Seeing may be believing but it is certainly not knowing. The light of consciousness streams out of our eyes and lights up what we think we see. When we go beyond sensory knowledge and penetrate into the depths of consciousness we find the real knowing, the real seeing. Only then will perceiver be united with what is perceived in the flash of illumination. Superseeing is seeing beyond the mental movie we make in our minds. We deceive ourself into thinking the movie is "out there" when really it is inside us, inside our consciousness. Real seeing involves shutting off the movie, to stop observing and start participating in the oneness that is All.

A NEW LANGUAGE

At the mental conceptual (blue) level of consciousness our ability to collectively confront and understand our world, each other and ourself is based primarily on the ability to name things in words by creating symbolic representations which capture and convey *meaning* by categorizing, relating, objectifying and binding into concepts our ever changing life experience. Symbols are the directly observable data of social intercourse and symbolic communication is a fundamental part of the dynamism of any society or social group. Psychologist Dr. Roberto Assagioli notes that "symbols as accumulators, transformers and conductors of psychological energies have most important therapeutic and educational functions." (1) Symbols give shared meaning to the humor and pathos of the drama of social living. However, we must also look beyond the symbols at the reality they stand for and communicate beyond the words.

Symbols in the form of words are often barriers to knowing the truth because learning to speak a language means learning to think in terms of abstract verbal concepts and grammatical categories. John White comments that:

> Language is symbolic: language learning is symbol thinking. As such, language constricts consciousness and places limits on understanding. One must go beyond words to a direct, unmediated vision in which perceptions are not filtered through a linguistic screen existing in the mind. (25)

Images, especially holographic ones, can compress a great deal of information into a very small space ("A picture is worth a thousand words.") And for the sake of our sanity it appears that compression of information into significant symbolic images which reach into the nonverbal depths of human life is the next step in our evolution together toward planetary consciousness.

The development of a transcultural species-wide language of images has always been gradually evolving but now is speeding up as the audio-video capabilities of supersensonics expand. This new language of direct perception dissolves the illusory barriers that separate us and helps to unite us for it is synergetic (more than the sum of its parts), syncretistic (combines many languages into one) and synaesthetically unifying (brings the peace of wholistic vision by joining harmonic opposites together).

All these new inventions, imaginative visions, new art forms and individual and group efforts when used toward a species-wide experience of transcendent vision will accelerate the curve of expanding consciousness in the human species as a whole. This will come about only if the avatars of synthesis come forward to make these visions realities. If the technology remains in the hands of those interested in profits and power we are in for much more of the inane banter that passes for communication today. A change in consciousness is the first step if the world culture is to grow and eventually embrace the entire Earth through all of us in a collective communal experience of evolving through self-consciousness, through group consciousness to planetary consciousness and beyond to the final, undifferentiated level – Cosmic or God Consciousness.

PSYCHOSOCIOLOGY

In the social sciences, studies at the macro level of world society tend to focus on large social "units" (such as nation states) which are believed to shape the destiny of smaller component units. This approach on the part of sociologists and social psychologists relegates the most fundamental "unit" of society, the individual, to the position of a relatively passive and ignored nobody who is acted upon and molded by "social forces." This results in proposals for social engineering presented to governing bodies which are further and further out of touch with reality. This is dehumanization and oppression of the worst sort for it leaves out or denigrates human love, purpose, intent and individual freedom of choice. It elevates outward behavior and experience above inner conviction, human feelings and conscious experience.* Charles Hampden-Turner in his deeply insightful critique of contemporary sociology, *Radical Man,* notes that: "we know from our everyday social judgements we make of people that their external appearances are not only misleading, but often trivial compared with deeper understanding of their character." (6)

He delves into the deeper psychological determinants of social behavior and presents a model of society as a psychosocial development process. He cites the hidden assumptions, experimenter biases and subtle tricks and deceptions which abound in contemporary sociological and psychological experiments which cast great doubt on their results. These studies reach such high levels of abstraction that only specialists in the field can understand them. A science about human beings should be comprehensible to human beings if it is to heal, nurture and enlighten us.

Psychosociologists need to study the powerfully transformative effects of significant symbolic images coursing through the nervous system of planetary man. As sociologist Charles Baudoin has noted: "Every image has in itself a motor-drive ... images and mental pictures tend to produce the physical conditions and the external acts corresponding to them." Before we can do anything we must first have an idea, a self-created mental image of the action to be executed. Imagination comes before action and energy follows thought. We change as our mental self-image changes. The developments in supersensonics are bound to bring changes in self-images on a mass scale, for they will increasingly challenge and confront everything we assume to be real.

---

* For example, instead of treating the bodies of school children with drugs to control them from without, we need to find ways to help them develop self-control from within by returning spiritual teaching to the classroom in the form of nonsectarian techniques of centering and meditation to awaken the intuitive and imaginative levels. "Meditating With Children" and "Meditation For Children" by Dr. Deborah Rozman are being used by many counselors, teachers and school systems for these purposes with great success. These books are available from the publishers.

One of the most important characteristics of any society is its vision of itself and its future, what sociologist Kenneth Boulding calls "organizing images." It is the image we have of the future which is the key to that future coming into realization. "Every society has an image of the future which is its real dynamic." (18) As our individual self-images change, our vision of the future changes. As a new image of man as a planetary being unfolds in each of us we expand our self-image to the group level, join the family of humanity and become world citizens. But to do this we need a model to follow, one that is based on the laws of nature and promises success because it is in accord with the cosmic scheme of things, the divine plan, the guiding logos.

NATURE'S MODEL OF THE NUCLEUS: THE LIVING CENTER

As we penetrate into the depths of atomic structure definite energy patterns of behavior in the world of the nucleus are revealed to our expanded vision. These behavior patterns have their corresponding hierarchical identity at the cellular, human, species and galactic levels ("As above, so below."). Through the centering, energizing and consciousness-expanding exercises of meditation we learn to bring together and integrate our personality so that it becomes resonant with our nuclear Self. This lets the light of consciousness shine through us clearly and purely, raising every atom and cell in our entire being to a higher energy state and bringing a higher and more integrated level of conscious self-awareness.

A similar process can occur at the group level when the group members are consciously aware of and diligently practicing the self-organizing and self-transforming techniques of nuclear evolution. This dynamic model of the process of increasing awareness takes place in every life form, the growth from self-consciousness to group consciousness.* Just as particle-waves join together to form an atomic nucleus with a positive charge that is balanced by the negative charge of the orbiting electrons, so can a group nucleus of people be formed which exhibits similar properties. In an interpretation of nuclear evolution, Deborah Rozman notes that:

> In a group of people, a true group nucleus is formed when individuals joining a group resonate in such a way that they "merge"* just as does a molecule of water when it hits a drop of water; they become selfless and annihilate their separateness. Universal awareness, universal love, then manifest through their being. The working out of this process of love is the evolutionary growth of God's expression of Love. Its operant method is through Principles of Radiative Unity. (31)

* For a deeper analysis see "Nuclear Evolution" (8) and "Hills' Theory of Consciousness." (15)

The particle-waves (members) of an atomic (group) nucleus when vibrating in resonant, harmonic unison (expressing the light of living love) tune the nucleus to receive incoming particle-waves (potential new members). This receptivity is based on the exclusion principle.

> Nothing can be absorbed into a nucleus until it is ripe or until a place has been made for it. The exclusion principle says that no more than one electron can occupy the same orbit around the nucleus at the same time. (31)

Lower energy (level of consciousness) newcomers enter and leave with little effect on the nucleus. But particle-waves (aware beings) vibrating at a more intense, higher frequency are attracted into the nucleus and cause it to change to a higher energy state. This is how atoms, cells, people and groups grow.

The constituent members of the nucleus reinforce each other with reflected waves (the group interaction as they think and act more as a whole) creating a resonant standing wave in them all. This wave enables the nucleus to absorb energy of a much higher frequency and at a much faster rate (logarithmically and exponentially) than the sum total of the individual members. This energy lies in the higher octaves of the spectral hierarchy of the light of consciousness, including cosmic rays and beyond. This is how atoms, cells, people and groups absorb life energy.

As members of the nucleus work on increasing the life energy flowing into their atoms and cells through the techniques of nuclear evolution, this light absorbed from the group standing wave is consciously radiated, just as the nucleus of a radioactive atom emits X rays, gamma rays, light, etc. This lifts the entire group to ever higher, more integrated states of consciousness, eventually bringing about true group consciousness through the positive, cohesive force of radiated love. Recalling our discussion in Section One, when cells in our body containing phosphorous atoms get supercharged with re-emitted energy, our bio-energetic field becomes brighter and merges with the fields of others as revealed by electrophotography. This produces a glowing halo in the human aura which is slightly beyond the visible spectrum, yet is quite clear to anyone who has conditioned his or her mind and visual pigment to see it (through the techniques given earlier, the course *Into Meditation Now,* the Rumf Roomph Yoga series and other exercises available from the publishers).

As the standing wave in the group nucleus increases in frequency and amplitude, there is a heightened phosphorescence in the cell life of each member as it absorbs and re-emits more life energy and cosmic radiation, eventually learning to continuously generate this light from within even when the exciting agent (the group standing wave) is removed (just like a firefly generates its own light). Eventually this increasing bioluminescence gradually comes into the visible light spectrum bringing to normal human vision the transcendent radiance of the *TRANSFIGURATION*. When all of Christ's disciples received the Holy Ghost at Pentecost they became a living example of this principle of radiative unity.

> True unity and harmony occur when the units of consciousness come together in the domain of self-surrender. Then a truly evolved nuclear evolutionary group is created birthing a new social organism as an evolutionary force for God-willed change which is the externalisation of the Kingdom of Heaven on Earth. This is the intent and real meaning of the first two or three lines of the Lord's Prayer. (31)

THE POSITIVE CENTER PRINCIPLE

At the center of the nucleus of every atom there is a strong positive charge, a will to unite, which is opposite in polarity to the negative surface tension which causes separation. The positive charge attracts the nuclear binding or cohesive love energy which acts as a solvent to penetrate the thin membrane which separates the individual particle-waves and unites them, bringing the selflessness of unitary consciousness.

Before this selflessness can manifest, the negative repulsive tension, a disruptive, conflicting force, has to be overcome. This force is a natural electromagnetic effect and is creative in that confronting and changing it brings growth.

> That we grow and evolve to love through conflict is part of the electrical dynamic of nature. The handling of the creative conflict process can be optimized through conscious creative conflict sessions, part of what might be called social nuclear evolutionary methods. The success of these sessions depends on a positive will response that provides the energy for the activation of the positive charge at the center to make nuclear evolution happen. (31)

This positive will response is created by commitment. The depth and manifestation of commitment is directly proportional to the strength of the positive charge. When that commitment is total no separation is experienced between self-will and the Divine Will. The gap between the manifested level of consciousness and the potential God-freed state is narrowed and finally bridged.

CREATIVE CONFLICT

The creative conflict group is committed to total openness. This commitment to participate and open up gradually ends the illusory self-sense, the detached ego, the observer consciousness which separates us from each other and our true Self. This is a prerequisite to achieving the direct perception of center consciousness, the ability to see situations and life experiences as they really are from a greater perspective, to perceive purely. This is the key to self-realization and the transmutation of energy (consciousness).

Harmonizing the conflicting energies of people and making them creative requires open communication based on tested principles that really work. If there is something manifesting in our consciousness that disturbs us, then that something must be confronted and mastered if we are to be at peace and to grow. This is the first principle of creative conflict. *We are whatever disturbs us.* For the purpose of making conflict creative we agree to sit down together and disagree openly in order to get to the real cause of the disturbance. This is where a group is helpful, since the situation can be viewed from several perspectives and the group mind and spirit can focus beyond personal emotions and attachments.

To communicate openly real listening is required. Not just hearing the words but going beyond them to the deeper meaning, the real essence that the person is trying to express. The technique of mirroring is especially helpful here. Mirroring is when we feed back what another person has told us, trying to express the same gestures and inner mood, whether or not we agree with them. This technique is based on a second principle: *we agree that everyone has truth from their point of view.* Mirroring allows us to get inside the inner world of another and hear what their being is really saying to us. It shows them whether we are really listening, or have put our own truth ahead of theirs by filtering what they say to suit our own perception of the situation. This is called projection, when we see something in others that is more in ourself than in them and colors what we see. The mirroring technique helps to eliminate projection and makes us more aware, deeper listeners because at any time we can be asked to mirror back what was actually communicated. When practiced continuously it reveals gaps in communication that most of us are never aware of.

The *red herring principle* is sometimes used to keep the discussion from wandering. This means that we do not change the subject, ask irrelevant questions or throw out evasive answers. Instead, we examine our own motives for disagreement before we doubt the statement of another. We look for the basic assumptions in what is said and find out for ourself if it is true or not. We also *refrain from quibbling* because we are committed to seeking the conflict in each person and the contradictions in their minds. We do not quibble over the words and ideas that are expressed because they are that person's truth (the second principle).

Finally there is the principle of *refraining from making a speech*. That is, we look to our own life experience as our authority rather than quoting books or what someone else said. In the same vein we do not make untested suggestions for others to carry out. Instead we empathize, putting ourself in the other's center, remaining serene and calm while listening to any bullshit, tantrums and barbs that may emerge.

Along with these principles we use certain communication techniques which help us get in touch with what is disturbing us. At the beginning psychodrama, originated by Moreno in the 1920's, serves many purposes in expressing inner conflicts. It helps us get in touch with our deeper self, gain confidence and ability to express ourself, our feelings and our innermost thoughts rather than repressing them. Emotional release and open expression is necessary before real sharing and creative resolution of conflict is possible.

The principles of creative conflict and the techniques of psychodrama help us get in touch with all seven levels of consciousness. Physically (red) we confront and act out real life situations, making them now. Socially (orange) we interact and express ourselves together. Intellectually (yellow) we analyse what's going on to penetrate deeper into the meaning and find solutions to problems. The process is full of life energy and love (green) that grows as we work together on our problems of insecurity, attachment, loneliness,etc. and build trust. Mentally (blue) we wrap our mind around the situation and get in touch with memories of past situations in order to cite specific examples, find patterns and help ourselves to change them by changing ourself. This is real growth. It is an intuitive (indigo) exercise for we train ourselves to tune in to other people's thoughts, words and feelings, into their real needs and life situation. It is imaginative (violet) because we see our life before us, like a dream made real. It gives us a chance to stop time, go back and see how we could have done things better and go forward by seeing what we need to change. This process gradually takes us beyond personalities to the oneness and attunement that develop as we progress and see the potential for humanity -- our planetary being -- coming together as one great family, as we come together in small creative conflict groups.

These techniques of open communication can be used as the basis for negotiation between the most divergent individuals and groups, from labor/management contracts to agreements among national leaders to international disputes at the United Nations. It is a direct way to expand awareness of each other's problems and needs and find solutions, progressively embracing more and more of us in yet higher levels of nuclear evolution to the planetary level – and beyond. Creative conflict is practiced daily by the student/faculty of the University of the Trees, a consciousness research center nestled among the towering redwoods of Boulder Creek, California. There is a tape available, Rumf Roomph Yoga number 16, that explains creative conflict in more detail. For further information write to the publishers.

WORLD GOVERNMENT

The study of world politics is increasingly becoming the study of the psychological communication processes at work in the world culture of world society, psychopolitics.* As political scientist Dr. Ralph Pettman notes:

> The real nature of mankind's advance lies in the development of culture. Culture has become one of our chief adaptive mechanisms .... Social communication is the plastic medium on which the transmission of culture depends. (20)

Drs. Lucian Pye and Sydney Verba place great emphasis on the communication of values, beliefs and expressive symbol systems and their effects on the political aspects of culture in political development (evolution). (21) A crucial concept in their view is that of *identity*. Many problems in political evolution in any society revolve around personal feelings about and ideas of belonging. Distinguished psychologist Dr. Roberto Assagioli has recognized a fundamental psychological principle which he has formulated as follows: "We are dominated by everything with which our self becomes identified. We can dominate and control everything from which we dis-identify ourselves." (1) There appears to be an identity crisis occurring throughout the world as more and more of us question who we are and where we belong in the cosmic scheme of things. The whole world is, people everywhere are, in a state of intense psychological transition at this time as we begin to realize that the state is a state of mind and that the state of consciousness changes when we change our concepts about it in our mind.

As long as we identify ourself first and foremost as part of some abstract group, for example as a citizen of a nation-state, we risk getting mentally trapped and emotionally embroiled in dichotomous "we--they" (we good, they bad) thinking. This is the kind of sensationalism and false separation that causes wars and has brought us to the critical impasse where the whole world is an armed camp and we stand at the brink facing nuclear annihilation together. It is time we stopped this ridiculous rhetoric, this war of words, which leads to the politics of hate, of alienation (separation) and potential violent conflict. The international system

---

* The science of psychopolitics is presently preoccupied with the biographies and personalities of famous political leaders from the perspectives of various "schools" of psychology. When the fundamental importance of consciousness and the psychophysiology of nuclear evolution is realized, this science will expand and become more exact and closer to reality.

of nation states only perpetuates this kind of separation by placing the sovereignty of that new god on Earth, the state, above the sovereignty and freedom of the individual. As Christopher Hills notes: "the rights of states are growing everywhere and the rights of men are lessening, because a sovereign Democracy does not practice outside its own domestic sphere anything but autocracy."

We need to return to the politics of love, of brotherhood and of *creative* conflict. It is time to realize that we are first and foremost human beings. Though we speak different languages and have different anatomical charcteristics such as skin color, we are in essence one family, one being. It is all "us" now. We are all in this evolving process called life together. There is no "them" anymore. We are all interdependent in the world society. A late frost in Brazil causes a coffee shortage and prices skyrocket. An oil embargo threatens to cripple Western industrial civilization. Shortages of vital resources bring international intrigues and political machinations that topple governments, cause wars and terrorist movements and increase world tension as the spectre of nuclear holocaust hangs daily over our heads. Political philosophers and psycho-political scholars everywhere are decrying the crisis of values, dehumanization and pollution of mind characteristic of twentieth century separated man. Rather than a cause for alarm or dismay, however, these world tensions should be seen for what they are: part of a divine plan leading to the coming new identity of planetary being.

Our society is unique in that it is the first known world society. As more of us identify at the planetary level we evolve toward planetary consciousness (the ecology movement is a good example). The barriers to deeper communication fall as we realize our inherent oneness. This evolutionary process of expanding world political awareness is unwittingly aided by the multi-mass-media communication systems. As Youngblood observes:

> Global information is the enemy of local government, for it reveals the true context in which that government is operating....Television makes it impossible to maintain the illusion of sovereignty and separation which is essential to their existence. (27)

More and more of our problems (the population explosion, the energy crisis and the pollution crisis, for example) are planetary in scope and require planetary solutions. These problems are generating pressure for greater cooperation and some kind of world government with a constitution that protects the sovereignty of the individual, and individual freedom is nurtured to develop the full creative potential of every human being. Freedom economically, politically and intellectually can only be derived from a deep psychological awareness of our life circumstances and our real needs and aspirations (which nuclear evolution provides). As Pettman notes:

> Economic freedom would mean freedom from the economy -- from being controlled by economic forces and relationships; freedom from the daily struggle for existence, from earning a living. Political freedom would mean liberation of individuals from politics over which they have no effective control. Similarly, intellectual freedom would mean the restoration of individual thought now absorbed by mass communications and indoctrination, abolition of public opinion together with its makers. (20)

We can have these freedoms, if we choose to have them.

MANNA FROM HEAVEN

Solar energy can be harnessed to solve the world food crisis, the world energy crisis and the world pollution crisis in one stroke. A solution lies in nature's basic link in both the food chain and the carbon deposits of coal and oil -- microalgae.

A specially developed strain of microalgae called spirulina (after its spiral shape) was discovered in Africa and adapted to different climatic and temperature zones so that it can be grown in many parts of the world. (30) It grows at prodigious rates with just water, fertilizer and sunlight. Under ideal conditions it multiplies logarithmically at the rate of 40 times in 24 hours. One ton of culture then, becomes 40 tons after one day, 1600 tons (40 X 40) after the second day, 64,000 tons (40 X 40 X 40) after the third day, 2,560,000 tons (40 X 40 X 40 X 40) after the fourth day, 102,400,000 tons after the fifth day, and 4,096,000,000 tons after the sixth day (one ton for every man, woman and child on the entire planet in just six days). On the seventh day we can rest.

Even under less than ideal conditions the production rate is extremely high, dependent only on the size of the production area, the necessary water, pumping, filtering equipment and fertilizers to promote fast growth.

The finished product can be burned in place of coal or wood. It can be processed for conversion into many of the chemicals that are now derived from petroleum. By adding preservatives, it can be stored for years. It can even be compressed and used as a wood substitute for building materials, thus solving the housing problem and the problem of storing vast amounts of surplus protein and fuel in times of abundance as insurance against famine or energy shortages. As a cheap, abundant and renewable source of energy it is unrivalled.

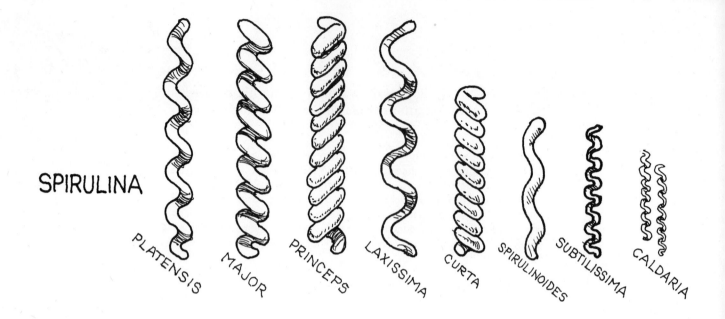

Microalgae can also be used as food. Through special processing it can be made pleasing to our eyes and through natural flavorings pleasing to our palate as well (look what food processors are doing with soy protein right now). As a food it is a natural source of many vitamins and minerals and as a source of protein it is unsurpassed.

COMPARISON OF PROTEIN CONTENT WITH COMMON FOODS
(PER CENT IN DRY WEIGHT) (30)

Spirulina	62 -- 68
Soy Bean	39
Beef	18 -- 20
Egg	18
Fish	16 -- 18
Wheat	6 -- 10
Rice	7

With the sun's help (and artificial light for 24-hour production) spirulina microalgae can convert many different fertilizers into chlorophyll, protein and hydrocarbons. It can live on special chemicals derived from petroleum (expensive), organic wastes such as chicken and cow manure, denatured human excreta, by-products of combustion processes such as engine exhaust fumes, etc. (cheap). In fact, the lakes and streams which are presently clogged and polluted by nonbiodegradeable phosphates, oil and other pollutants can be cleaned by establishing microalgae cultures, balancing the phosphate/nitrogen constitutents in the water and algae and pumping out (harvesting) the algae. This process not only cleans up pollution but also directly converts it into an excellent energy source. Even if we extract the inedible wild algae from lakes we can get millions of tons of burnable fuels and chemical derivatives. Sewage plants can serve as sources of fertilizer to provide millions of tons of dried fuel annually.

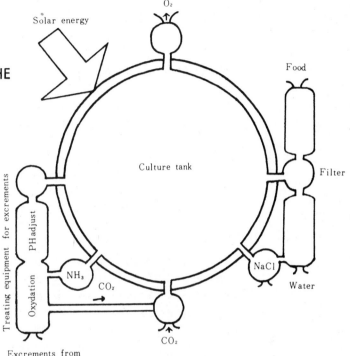

CLOSED CULTURE SYSTEM FOR THE HYPERACTIVE PRODUCTION OF SPIRULINA ALGAE

Remarks:
(a) The tank is a gas-exchanger ($CO_2 \rightarrow O_2$) as well as the culture of algae.
(b) The culture tank made of a transparent body in globular shape rotates slowly while absorbing solar energy from the window.
(c) For use of excrements from the human body, oxydation in a wet-system is desirable.

With another microalgae, chlorella, the necessary technology for large scale production (using a centrifuge which can separate 90 tons of product per hour) and harvesting by precipitation techniques has been perfected over the past fifteen years by over 150 experts in the Microalgae International Union (MIU).* At this time there are microalgae projects underway in Japan, China, Taiwan, Africa and the United States.

The only reason that this technology has not been adopted on a massive scale to solve the energy, food and pollution crises is because of another crisis, the cooperation crisis. Representatives from MIU have met with increasing frustration in trying to get governmental financial backing for further research and production equipment. Vested interests in soil-based agricultural methods have blocked attempts for government funding -- while people starve. Energy prices keep going up when there is

---

* Formed in 1964. President is Dr. Christopher Hills. Information on microalgae production is available from the University of the Trees, which also offers independent study courses on the subject which are creditable toward a degree.

a plentiful source of energy available right now. Government on the basis of the self-interests of exploitative financial groups may eventually starve or fry us all, unless we wake up and develop our political and intellectual freedom. Just as microalgae is a potential solution to several economic crises at once, there is a potential solution to the present political impasse and feelings of powerlessness and alienation that many of us feel today.

THE PEACEFUL CENTER

Breakthroughs in the form of new methods of self-government will come by increasing our bioenergetic and psychoenergetic sensing abilities. These depend on our capacity to actively use the power of internal vision (imagination) to condition ourself rather than passively allowing our surrounding environment to condition us. Effective self-government is self-healing self-help which comes from recognizing the source of all wisdom and all order through law, the peaceful center deep inside where we are all one.

To point a finger and tell or try to force others to be peaceful is an exercise in futility. In creative conflict we begin where we can be sure of success, with ourself. Love is the cohesive, all embracing life energy which moves all creation. Love brought us into existence, nurtures us through it and returns us beyond it. Without love, we are nothing at all. The "dark night of the soul" and the journey through the "valley of the shadow of death" are allegories for our penetration into the depths of our heart and, hence, into the heart or central source of us all. Wisdom and compassion, head (diversity) and heart (unity), are the two primary attributes of peaceful enlightened beingness. How can we give these to others unless we have them in ourself? If we truly want peace in our world, then we obviously have to work arduously and unceasingly at bringing that peaceful center into existence at the heart of our being and then radiating that peace into other centers through the light of living love.

A CONSTITUTION FOR WORLD PEACE

Peace researchers write volumes of complex and erudite theories about conflict resolution, nuclear disarmament, a world police force, ad infinitum as ways to world peace, yet most of them miss the essential point that to eliminate world tension we need to reduce individual tension, beginning with ourself. Individual tensions aggregate into world tensions and have their origins in frustrated drives (the love drives of nuclear evolution) which lead to anger, hostility and assault (war). Conflict will never be resolved through untried theories taught in schools, nor can peace be enforced through threats of punishment by violent means. Violence begets violence and these supposed ways to peace are counterproductive and by increasing tension actually contribute to the destructive conflict they are meant to resolve. World peace will only come when all of us find peace within ourself through creative conflict which channels conflicting energies into growth producing activities. To bring that peace we need new forms of self-government dedicated to peace. They must be in harmony with the laws of nature, be based upon unanimous consent, foster human love through community sharing and encourage the fullest development of each person's creative potential.

The Center Constitution outlined in the new edition of *Universal Government by Nature's Laws* (available from the publishers, Spring, 1978) embodies these principles. This is a model constitution based on the sovereignty of the individual, not the state. It is a creative and dynamic, evolutionary legal document for bringing inner peace to individuals and social groups which can eventually be applied as a practical model for a world government. The Center Constitution has also been adapted to small communities numbering at least 12 members so that it can be tested in real life experience to yield practical results and undergo the evolutionary changes to make it suitable for all. Only when such constitutional experiments have been proven to work will they communicate a new existence to humanity through those of us who work the new system. The Center Constitution encourages and brings about new art forms through the lively enthusiasm and creative imagination of the participants. Only by working it in small groups first will the world be ready for it and desire it. Until the world is ready small groups of sincerely dedicated people like the Centre Community in London, England, and the University of the Trees in Boulder Creek, California, are evolving this constitution as we evolve toward planetary and cosmic consciousness in community peace centers. These small groups can be started by anyone and offer the quickest growth by solving inner and outer conflicts.

THE WORLD PEACE CENTER

The groups working the Center Constitution are building a framework for self-governing communities of world citizens who will eventually form a model World Peace Center. For it to have any real meaning on a planetary scale, a world constitution must obviously be written by people of all cultures living together. The purpose of this center, therefore, will be to bring the diverse cultures of Earth together where a lively and loving cross-fertilization of hearts and minds will nurture an ethic of human fellowship. This close family spirit will give these pioneering individuals a chance to exercise the power of unity for social and spiritual ends. Through the tests of real life experience the Center Constitution and the World Peace Center will evolve as we evolve our ever expanding awareness.

But all the finest words and best intentions are worthless without action. The world knows that action speaks loudest. The ideal becomes real when we put theory into practice. No one can do it for us. Experience of life is still the best teacher and dynamic growth is the result when we participate in life. If you really want to be a part of a grand experiment in the divinity of Man, let us hear from you. We all need your help, for we are you and you are we and together we must unite to conquer the forces of division.

Some big surprises may be in store for us when we better understand each other and relate more intimately with the energy and form in the environment. The scientific method alone can never lead to direct perception of reality because it requires us to use external instruments. When we become the instrument of direct perception we will be able to see and know the deepest secrets of matter in the worlds around and inside every being. Direct perception unites pure science with pure spirit in us as we seek union with the divine Absolute, the highest state of *PURE CONSCIOUSNESS* in the *ONE*.

The world is a mirror of our real Self. It gives us instant, often painful feedback on where we are really at. A smoky mirror gives only a dim reflection of a hazy image. But a clean and shiny one reflects us as we really are. To see clearly, to perceive purely we have to get up off our comfortable seats and polish our mirrors constantly so that we can better reflect the Truth to one another. The World Peace Center will hasten this polishing process on a planetary scale.

Look into a mirror.

What do you see?

Who is that someone,

Looking at me?

Invisible, untouchable,

Unknowable am I.

That ONE who sees

Through every eye.

REFERENCES FOR SECTION TWO

BOOKS

1 Assagioli, Roberto. Psychosynthesis. New York: Viking Press, 1965.

2 Beasley, Victor R. Dimensions of Electro-Vibratory Phenomena. Boulder Creek, CA: University of the Trees Press, 1975.

3 Capra, Fritjof. The Tao of Physics. Berkeley, CA: Shambhala, 1975.

4 Dass, Baba Ram. The Only Dance There Is. Garden City, New York: Anchor/Doubleday, 1974.

5 De Chardin, Pierre Teilhard. The Phenomenon of Man. Translated by Bernard Wall. New York: Harper and Brothers, 1959.

6 Hampden-Turner, Charles. Radical Man: The Process of Psychosocial Development. Garden City, New York: Doubleday, 1970.

7 Hills, Christopher. Supersensonic Instruments of Knowing. Boulder Creek, CA: University of the Trees Press, 1975.

8 Hills, Christopher. Nuclear Evolution: Discovery of the Rainbow Body (second edition). Boulder Creek, CA: University of the Trees Press, 1977.

9 Hills, Christopher. Rays From the Capstone: Harnessing Pyramid Energy. Boulder Creek, CA: University of the Trees Press, 1976.

10 Hills, Christopher. Supersensonics: The Spiritual Physics of all Vibrations from Zero to Infinity. Boulder Creek, CA: University of the Trees Press, 1975.

11 Hills, Christopher. Universal Government by Nature's Laws. Boulder Creek, CA: University of the Trees Press, second edition to be published Spring, 1978.

12 Kaufmann, William J., III. Relativity and Cosmology. New York: Harper and Row, 1973.

13 Mann, Felix. Acupuncture: Cure of Many Diseases. Boston: Tao Books, 1972.

14 Massy, Robert. Alive to the Universe: Handbook for Supersensitive Living. Boulder Creek, CA: University of the Trees Press, 1976.

15 Massy, Robert. Hills' Theory of Consciousness: An Interpretation of Nuclear Evolution. Boulder Creek, CA: University of the Trees Press, 1976.

16 Mishlove, Jeffry. Roots of Consciousness. New York: Random House, 1975.

17 Moss, Thelma. The Probability of the Impossible. Los Angeles: J.P. Tarcher, 1974.

18 Muses, Charles and Young, Arthur. Consciousness and Reality: The Human Pivot Point. New York: Avon Books, 1972.

19 Ostrander, Sheila and Schroeder, Lynn. Handbook of Psychic Discoveries. New York: Berkley Publishing Corp., 1974.

20 Pettman, Ralph. Human Behavior and World Politics. New York: St. Martin's Press. 1975.

21 Pye, Lucian W. and Verba, Sydney. Aspects of Political Development. Boston: Brown, Little and Co., 1966.

22 Puharich, Andrijah. Beyond Telepathy. Garden City, New York: Anchor/Doubleday, 1973.

23 Rhine, Louisa E. Mind Over Matter. New York: McMillan, 1970.

24 Toben, Bob. Space-Time and Beyond: Toward an Explanation of the Unexplainable. In conversation with physicists Jack Sarfatti and Fred Wolf. New York: E.P. Dutton and Co., 1975.

25 White, John (ed.) The Highest State of Consciousness. Garden City, New York: Doubleday, 1972.

26 Wood, Ernest. Concentration: An Approach to Meditation. Wheaton, Illinois: Quest Books, 1949.

27 Youngblood, Gene. Expanded Cinema. New York: E.P. Dutton, 1970.

ARTICLES AND PAPERS

28 Delgoff, Eugene. "A Holographic Brain Model," City College of New York, Date unknown.

29 Hameroff, Stuart Roy. "C'hi; A Neural Hologram? Microtubules, Bioholography, and Acupuncture," American Journal of Chinese Medicine, Vol. 2, No. 2 pp. 163-170, 1974.

30 Hills, Christopher. "The Mass Production of SPIRULINA, a Helical Blue-Green Algae as a New Food." Boulder Creek, CA: University of the Trees Press, 1974.

31 Rozman, Deborah. "Manifesting Human Potential States in Social Groups Committed to Actualization," Boulder Creek, CA: University of the Trees Press, 1974.

32 "Hear the Light," public announcement by Bell Laboratories/Western Electric, 1977.

33 "Hitachi Develops Holographic Videodisc System," Videocassette and CATV Newsletter, December, 1976.

SECTION THREE

VIBRATION AND FORM

INTRODUCTION

If someone came up and asked us where we live the odds are very good that we would start to describe a group of buildings, trees, rooms, people and streets, in other words a group of forms. Most of us live in a world made of forms: visual forms, audio forms and emotional forms are all around us as we go about the daily business of living our lives.

One of the most exciting developments in recent years is the growing acknowledgement among people from all branches of science that the forms that seem to make up reality are personal realities unfolding like a movie before our mind's eye, different for each of us yet at the same time similar enough to share many characteristics. Each of us has the power to create our own forms of reality, to shape the images with which we experience the world so as to make our lives full of joy or replete with horrible accidents.

As we have seen in the previous sections, these forms are actually vibratory patterns, constantly changing energy relationships which only appear to us as solids because our senses cannot detect the vast spaces between the subatomic particles. Consciousness is intricately related to our world of matter, and the phenomena of telepathy, psychokinesis and biological transformations due to psychic influences show that this invisible stuff inside us not only affects material reality but also creates new material circumstances as well.

The role of consciousness is taken for granted in all of our technological creations and all the credit is given to science and progress even though scientific discoveries are all made with consciousness. The technical marvel of the suspension bridge is first and foremost a creation in the consciousness of the engineer who conceived it and only secondly a solid material form. Yet our world pays all homage to the secondary form and the consciousness within the engineer is all but forgotten, except by those other engineers who have to go and build a similar bridge!

Our material circumstances are reflections of our creative consciousness and what we think is the real thing -- all the material goodies such as cars, food, sex, boats and money -- is only happening inside our personal model of reality. The real show is something altogether different.

A look at our situation on this planet shows that most of us are caught in a conflict between the inner reality versus the outer reality. Our senses and biological instincts are caught up in an external world while the one who is experiencing that "external" world is floating in an inner world that exists in many other dimensions of reality. For many of us the relationship between the two is not clear at all and as a result we avoid looking at the conflicts (except when they surface from our unconscious via our dreams*).

This is the riddle of the Sphinx, a paradox of unlimited mind caught in the limited body of a beast. As we have seen in Sections One and Two a solution lies in acknowledging that these are different dimensions of the same being, that our outer world of sensations and our inner world of mentations are reflections of each other. What becomes important is the unified field in which mental forms and the material embodiments of these forms play their roles. The following chapters deal with new areas of research and experimentation which have been helping to outline the characteristics of such a unified field. We will see that the forms which make up our everyday reality are actually creations in our imaginations. The simple experiments which are included are designed to show us how.

* For a clearer exposition of this conflict and the unconscious mind see "Hills' Theory of Consciousness" by Robert Massy and "Nuclear Evolution" by Christopher Hills.

The electromagnetic spectrum

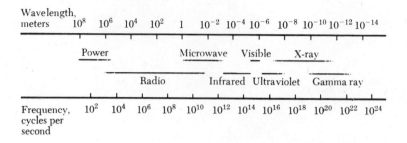

In Chapter Seven, The Voice of Creation, we will see how material forms and thought forms are created by vibrations. We will learn procedures which show not only how vibratory patterns are transformed into forms via denser mediums such as sand or water but also how our thought patterns can be used constructively for healthier and more fulfilling lives. Sounds and images can be used for programming new forms into our worlds of experience and by using the techniques of chanting a mantra as a springboard, we can begin to discover new frontiers of experience in the universe around us.

In Chapter Eight, Rays of Life, we will see how forms can act as lenses for focusing powerful vibratory fields by looking at the energetics of the pyramid form. The pyramid is a beautiful example of how powerful cosmic energies can be tapped by building a form. First we will investigate the actual shape and then move on to other forms which also act as powerful capacitors for the same energies: the Pi-ray Accumulator, the Positive Green Pendulum and hemispheres. Several experiments are included so we can learn how to work with the vitalizing properties of these energies as well as avoid the harmful effects that can result from their improper use.

After these preliminary excursions into ways in which vibrations create forms and forms focus energy, Chapter Nine -- Creative Imaging -- is concerned with the role of the investigator who is discovering differences between invisible vibratory fields and visible forms. Many people are discovering that they can see without their eyes and proving that other skin cells can be trained to see. We do not need highly specialized eye cells in order to use the visual faculty of the mind because other unspecialized skin cells have proven to be photosensitive. In this chapter we will see that what we think we see with our physical eyes is actually occurring inside our mind's eye, which can use many different cells as sources of information. Although the eye needs to have a seer in order to see, the seer does not need an eye to see! Several experiments show how we can learn to see without eyes and begin to see with our mind's eye, how we can train the creative imagination to become sensitive to those fundamental vibrational processes causing the sensory phenomena.

Chapter Ten, Supersensing, introduces the built-in biological sensitivity to the vibratory fields emanating from all natural systems which every human being is born with. Taking what we have learned in preceding chapters about the vibrational nature of material forms as a starting point, we will see that the human nervous system is capable of tuning in to the vibratory pattern of anything in nature and distinguishing it from the ocean of other vibrations in which it is constantly bathed. This faculty, which is called by many different names such as supersensing, dowsing, divining, radiesthesia, and the biophysical effect, is dormant in most people yet with proper training can be awakened. Once this ability is developed it can take us on a voyage of discovery into the center of any vibratory system in nature. Some of the widespread uses to which this faculty can be put include prospecting, detecting thoughts from others, investigating chakra dynamics (see Section One), mapping bioenergetic fields from atoms to stars, and determining the structure of our own consciousness.

Finally in Chapter Eleven, The Healer Within, we tie up this section on vibration and form with an introduction to radionics and energy broadcasting. In this chapter we investigate procedures for establishing counter waves which can neutralize the effects of disease-causing vibratory patterns and broadcast new healing energies into a person's wave field. As we increasingly discover that material situations and events are forms created out of vibratory patterns we are also finding that particular patterns which are causing us disease can be neutralized or equalized by using the principles of divining. We pick up this trail of discovery with the development of radionics. Diagnosis and treatment of illnesses with vibrational fields instead of material instruments such as drugs and knives is possible via fields generated by a radiomagnetic device known as a radionics set.

However, the main thrust of the chapter is not the radionics technologies but the exposition of the latest theory of energy broadcasting -- that it is all a function of consciousness. Results achieved through the use of sophisticated equipment have been duplicated by concentration and creative imagination techniques. This theory states that equipment is not really necessary but merely acts as a guide to help the operator concentrate and then imagine certain energy patterns which in turn lead to changes in the material structures and forms. All these findings are pointing the way to a new science of being in which we interact with nature as vibrational patterns within a vast unified field of vibrations. We are not separate entities acted upon by external forces beyond our control but participating members in a grand ballet in which we consciously create a beautiful work of art, our world community.

THE VOICE OF CREATION

Just as all material forms are actually condensations of high frequency light energy they are also the unique echoes of a primary source of all vibratory patterns. Pure light is invisible to the senses yet its reflections are seen everywhere as colors, shapes, outlines and intensities. In the same way the Voice of Creation is inaudible yet its echoes can be heard in every vibrating system in nature, not with our eardrums and cochlea but with the ear of the ear, the inner listener who makes sense of all the noise coming in through the ears.

As we will discover in succeeding pages, the key factor in apprehending our universe is the consciousness which uses the senses but is not limited by them. Even though the senses are severely limited we have other ways of knowing which transcend these limits and can soar to any part of the creation on the wings of the super sound current of the universe, the primary radiation of the Voice of Creation. These undeveloped ways of knowing are centered in the intuition and imagination and the training of these faculties is the first step in learning the new science of *supersensonics*.

VIBRATION AND FORM

To many of us the world we live in is one of stable forms and patterns and in fact we seem as a species to thrive on stability. The new physics however is rapidly shooting this world view down and perhaps by the end of the century it will no longer be taught. All around us is a world of seemingly stable patterns and forms yet when we investigate the fundamental nature of these forms at the subatomic level we find that they are constantly changing vibratory fields.

What is behind this puzzling phenomenon? Why do the forms we apprehend via our senses seem stable while we touch them with our hands yet when viewed from the physicist's perspective become clouds of changing vibratory fields?

The answer lies deep in the nature of our perceptual faculties which process the signals inputting through our sense receptors and present them to our awareness as forms. If we trace our perceptions back to their source we find that stable forms are actually psychic representations within our consciousness arising from the interactions between our vibrating consciousness and other vibrating fields of energy. These interactions result in interference patterns which are represented in our minds as images of stable material forms while in truth they are just bundles of vibrations like everything else. We then create an inner world populated with these imaginary forms and call it reality.

Ripples of water showing interference of waves in a tank, each having different wavelengths. When the wavelength is longer (left) the beams spread farther apart.

Many interference patterns are never represented within our awareness and given the status of reality because we favor the frequency range relating to our survival as biological organisms. Even though our consciousness operates over a much vaster range of vibrations than the senses can detect, most of us are not aware of this because we are overly preoccupied with biological level needs.

As our societies have evolved to the point where biological survival is no longer the full time job it used to be, we are becoming aware of other dimensions of our being which interact with vibratory frequencies operating in ranges on either side of that narrow slice occupied by our physical senses.

We are not separated from the total spectrum of frequencies and even though we are not consciously aware of it, many subtle interference patterns are at work in our consciousness creating patterns far subtler than those of physical matter, patterns such as ideas, emotions and images.

Can anyone touch an emotion or an idea? Our hands will go right through an idea or an emotion because they are vibrating in such different octaves that they do not interface with each other. This is the essence of the model of the seven levels of consciousness we introduced in Chapter Two. We are living in seven different vibratory dimensions, all interpenetrating each other yet not interfering with each other.

Actually we have the ability within our imaginations to tune ourselves to the total range of frequencies but we do not use it. Why? Because our preset biological survival tuning (food, shelter, sex etc.) is feeding us such strong signals that most of us do not want to change the channel. Essentially it means dropping our identities as limited bags of skin called human beings and not paying attention to the physical sensation level vibrations and this is the last thing many of us want to do. We like to make love, eat, touch, go to the movies and listen to music and are so wrapped up in these pastimes that the very thought of being without sensory stimulation is unpleasant to many of us.

Our personalities can be seen as the effects of subtle thought patterns in the medium of our actions. Even though the mind may not be matter in the sense that a brick is, it is a vibrational form, a subtler and more elusive form. This is what makes working with the mind such a difficult craft – its slippery and elusive substance is difficult to capture and mold into a design in tune with nature. Difficult, but not impossible. And when it is captured by a master mindsmith the work of art that is created is of a correspondingly deeper beauty.

VIBRATING FORMS

In order to substantiate our potential it becomes necessary to prove to ourselves with physical-sensational proofs that the world of matter is a secondary reflection of primary light vibrations, an echo of supersensonic sound. One day Ernest Chladni, an eighteenth century German physicist, clamped a steel disc to a table via a rod attached to its center, sprinkled some sand on the surface and then stroked the edge with a violin bow. He was utterly amazed to find that the frequency being played into the disc was creating a beautiful pattern in the sand.

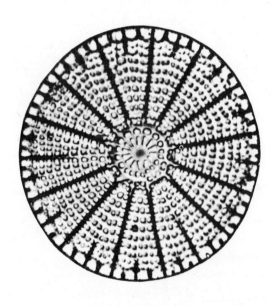

The bacterium Arachnoidiscus (x 600)

A Chladni figure formed by vibrating a sand covered disc to specific frequencies

Subsequent investigations showed that many patterns could be formed in mediums such as sand, water, oil and iron filings by subjecting them to wave patterns of sound. These findings were published in a book titled *Die Akustik* which laid the foundations for the science of acoustics and had a profound impact on the physicists of his day for they were instrumental in outlining how energy travels in wave fields and how interference patterns are formed.

All these findings were supposed to be facts about the "external" world and it took physicists another century and a half before they were face to face with the knowledge that there is no world of external matter, that what we see are the interference patterns created within our own vibrating consciousness.

OUR VIBRATING CONSCIOUSNESS

By seeing our life situation as a form created by the vibratory patterns within our consciousness, we can begin to see that we are creating the world in our imagination. In other words we can actually change the reality we experience physically, socially, intellectually, emotionally, conceptually, intuitively, and imaginatively by programming ourselves with vibratory patterns specifically designed to tune up specific areas of our being.

The techniques of *mantra* and *chanting* are particularly effective in this respect because, besides bringing new patterns of being into sharp focus within our consciousness, they also create new programs with physical sound vibrations and input them through our physiological receptors. Even if we do not get out of our sensation level experience of the world we can still input transforming patterns into these receptors which will eventually lead to supersensory levels of awareness.

A mantra is primarily a mental sound, as the words or sounds we hear are first conceived and energized by the concentrative faculty of the mind and then represented audibly via the larynx and voice box. The only real differences between the physical sound form and the mental thought form are wavelength, frequency and medium. The physical sound is a transformation of a pattern codified in mental wavelengths and frequencies into the slower frequencies and longer wavelengths of sound.

Using a mantra involves a form of thinking and reasoning in which positive solutions to problems are found; an exercise in creative thinking. In other words, mantras are tools for deprogramming and reprogramming our mental computers. Most of us are here on the planet because we have some problems to work out and because all of these problems arise from our mental tapes and habits it is important to learn how to erase these tapes from our bio-computers by using a mantra.

THE COSMIC WOMB

For instance if a man had a real problem with jealousy and could not stand it when someone else started talking to his girl friend, he could use a mantra to help deprogram his jealous reaction pattern and, by generating counter waves, substitute a sense of sharing in her new experiences. Instead of being imprisoned in his feelings of jealous possessiveness he could share in her freedom to be herself. In this case the mantra could be a phrase such as "I am whatever disturbs me." When he felt the old feeling creeping up and taking over he could say this phrase silently to himself as a method of getting in touch with the fact that his friend was not causing him any pain and the pain was his own doing.

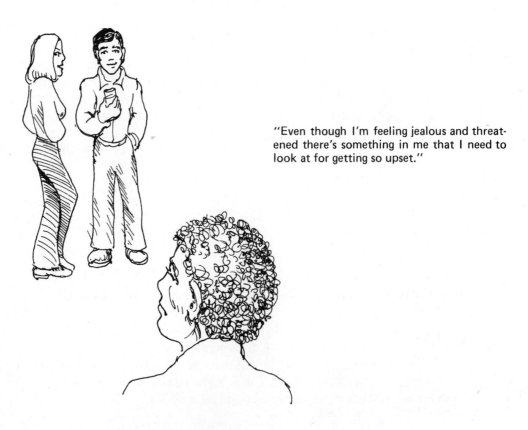

"Even though I'm feeling jealous and threatened there's something in me that I need to look at for getting so upset."

In the field of using mental vibrations to unlock the potential hidden within our seven levels of consciousness the ancient Indian yogis shine brilliantly. Thousands of years ago they discovered the basic unity between the light of consciousness underlying our perceptions and the pure light of which the cosmos is formed and called it *yoga,* yoga being the Sanskrit word for union. In ancient India most human beings were just as blind to yoga as we are today, so these sages created powerful methods for tearing away the veils of human awareness and expanding into cosmic awareness of yoga. Of these methods two of the most powerful are *mantra* and *chanting*. The ancient yogis discovered that sound could be used as a carrier wave of images that could help the student gradually open the doors of his/her perception. They developed a spiritual science of sound which is centuries ahead of our Western theories.

SOUNDS OF RE-BIRTH

Mantra is a Sanskrit compound word joining *man,* which means mind, and *tra,* which means both protection and instrument. Essentially a mantra is a tool for protecting our mind from getting trapped in false models of reality. The theory of mantra is that it is an image initiated deep in the core of our being that symbolizes a state of being we want to be reborn into. This image is then clothed in a pattern of sound that is symbolic of this rebirth, for example the Sanskrit phrase OM MANI PADME HUM, meaning Pure Consciousness is the Jewel in the Heart of the Form. Such a phrase can be used to symbolize a new way of seeing forms as temporary reflections of reality which will help the user to not get attached to possessing them.

The technique is to repeat the word throughout the day and especially when the old pattern is jumping up and down trying to take over again. A famous example of this is contained in the children's story about "The Little Train That Could". That little train is a metaphor for our inner being. As long as it thought to itself "I can't," there was no way it could climb the mountain (of self-discovery). Then it started to use the mantra "I Know I Can" and though the climb was steep it made it.

In the same sense the practitioners of positive thinking are using the principle of mantra when they substitute an optimistic and hopeful new image for the self-doubting image that is blocking success. As we will explore more extensively in Chapter Nine, success with mantras is tied in to the use of the creative imagination as a tool for transforming our experience of reality from drab existence to exciting discovery.

To be of practical value in our lives, the study of vibration and form must be incorporated into personal practices which somehow contribute to our inner growth and fulfillment – practices like the use of mantras and chanting. While there are literally thousands of mantras, some of the most effective are the ancient Sanskrit ones such as OM. The reason for this is that in Sanskrit, like Hebrew, each sound pattern corresponds directly to a fundamental pattern of nature whereas our modern words do not maintain this link.*

The word OM, for instance, is made up of the combined sounds of *A-U-M,* each of which expresses a primordial aspect of nature: *A* -- the subjective awareness of an external world, *U* -- the awareness of an inner being; and *M* -- the consciousness of undifferentiated unity with no conflict between an internal or an external. By repeating this sound we can tune ourselves consciously to these fundamentals of life itself, the accuracy of the tuning depending of course on the intensity of being that we put into making the sound. If the sound is repeated like a dumb gramophone with no feelings put into it, then the tuning will be weak because the tuner will not have enough power to really latch onto the signal. By really putting concentrated attention and feeling into making the sound and at the same time imagining the quality of life which we are trying to tune in to, the reception becomes stronger and clearer. In fact, as with radio circuits, when the tuning is precise, a resonance effect begins generating standing waves which blow the original problems away like so much hot air.

Whereas most words we use to represent our experience have lost this direct link with life processes, each sound does resonate with other vibratory patterns such as colors, crystals and elements, although these relations are usually below our thresholds of awareness. The beauty of the Sanskrit sound symbols is that, unlike English sound symbols, their connections to reality are not cluttered over with distorted meanings and consequently are more accurate. In terms of effectiveness the way in which we vocalize a mantra such as OM becomes important as well. This is why the technique of chanting was developed: for achieving deep and effective communication with the subtle vibratory patterns at work behind the sense-perceivable forms.

THE BREATH OF UNION

The word *shanty* comes to us from the sailors who voyaged to India and heard the Indian workers singing work songs with the words "OM SHANTI." These songs had such a power and lift to them that they soon

---

* The word individual is a good example of this breakdown in English. The two roots mean "not" and "divided" yet the compound is commonly acknowledged to mean "separate from." Such breakdowns are quite common in our language and vividly illustrate how our minds distort reality to the extent of arbitrarily changing the meanings of the symbolic representations we have devised for building our world models.

adopted them for making heavy work easier, calling them "sea shanties." Gradually the sailor's shanty became a chant and now we have the word "chant" in English or "chanson" in French.

Long ago, languages such as Sanskrit and Hebrew were chanted rather than spoken and even today when hearing these languages it is easy to sense that they were meant to be sung rather than spoken. Some clues as to why these ancient languages were designed to be sung can be gleaned from the differences experienced between chanting and regular talking. One is that while chanting is a continuous process, talking is made up of discrete units. Chanting is a flowing movement that encompasses and weaves the listener into itself, while talking is a series of still shots, fixing the listener into a spontaneous time world separate from that of the talker. Perhaps the most striking effect of chanting, however, is the deep feeling of inner calm that follows.

Physiological investigations into this phenomenon have shown it to be due to a special manner of breathing used in conjunction with the creative imagination, known as the Great Yogic Breath. This is a slow, deep breath that maintains a high pressure in the lungs and lets out the tiniest amount of air necessary to chant OM through the nostrils. This OM vibration in the nostrils causes an intense vibration in the sinuses and the four ventricles of the brain.

We can directly experience this effect anytime we wish: start slowly by inhaling into the stomach, then the lungs, then the throat, and finally the head. Hold the pressure and then slowly exhale in the reverse order. Repeat this exercise for five minutes.

Long, slow breathing makes us alkaline while rapid breathing makes us acid. The concentration of acid in our system determines our health, as over-acidity will burn up the cells. This acidity is a function of how many hydrogen ions (H+) we have in our tissue fluids and blood, and the concentration of these ions is directly related to our breathing of oxygen. Breathing oxygen produces (H+) ions and long, slow breathing increases the intake of oxygen into our blood through the lungs.

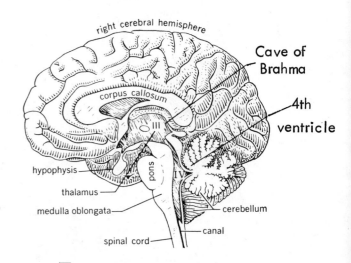

The diagram shows "The Cave of Brahma" at III referred to in the Sanskrit texts as the seat of resonance with the ONE ocean of cosmic vibrations. It manufactures the cerebrospinal fluid in IV which floats the whole brain and spinal nerves. The chemical Ph of the fluid determines the resonance.

Inside four cavities deep within our heads, known as the Four Ventricles, a liquid is produced - cerebrospinal fluid - which bathes the entire brain and spinal cord with rich proteins. The protein synthesis of this fluid is enhanced when there are more (H+) ions in the blood. Synthesis of proteins involves the formation of crystals from ionic substances in the cells, a process dependent on the bonding qualities of (H+). The importance of this fluid is that the condition of the proteins it contains directly determines our vitality. We are constantly absorbing powerful radiations from the cosmos through our chakras, our nerve plexuses and the liquid crystals within our cells. The translation of these energies into physiological vitality is mediated through these crystals, which are tuned to specific frequencies according to their angles of crystallization. By stepping up the protein synthesis in our cerebrospinal fluid we increase the number and variety of crystal receivers in the fluid and consequently the quality of the energies absorbed from the cosmos. In terms of psychic holograms we are expanding our capacity to radiate coherent light frequencies which can interact with incoming stellar light frequencies to form new patterns of being.

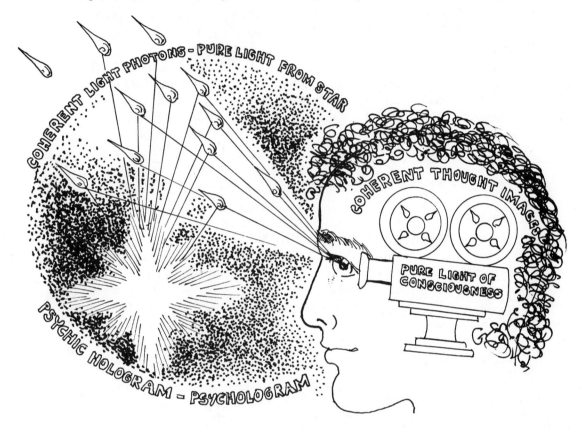

It is similar to the communication equipment in some whales whose heads are filled with tons of a special oil which acts as a resonator so they can communicate with other whales over long distances. The cavity in their heads and the special oil seem to act as a liquid crystal for transmitting and receiving particular frequencies, much like a radio set. When a sound such as OM is vibrating this fluid it is much easier to create a resonance with the cosmic forces symbolized in that sound. The object of this breathing method is to fully extract oxygen and excrete carbon dioxide. While fast breathing takes greater quantities of oxygen into the lungs, it does not lead to efficient extraction because the carbon dioxide is not being exchanged for the oxygen. The breath is so fast that most of the oxygen leaves just as fast as it comes in. Proper gas exchange needs the time and pressure provided by doing the Great Yogic Breath.

Another effect of the Great Yogic Breath technique is tranquility. Because the carbon dioxide is being exchanged for oxygen more efficiently, greater quantities of it accumulate in the air sacs of the lungs during exhalation. Carbon dioxide acts as nature's own tranquilizer doctors often use it for this purpose*— and leads to relief from stress, one of the major factors leading to over acidity. Slow and deep breathing also conquers nervousness because it enhances the acetylcholine concentration in the parasympathetic nervous system, thereby activating it strongly. This activation of the parasympathetic nervous system leads to greater relaxation as well as an increased sensitivity to extrasensitive telepathic communicating such as that practiced by whales. The ultimate aims of this exercise are to enrich the nutrients in our cerebrospinal fluid, activate our parasympathetic nervous system, alkalinize our body tissues, and promote good health.

HEALING WITH SOUND

The next step is to use chanting as a means to heal ourselves. *Heal* means to make whole and anything we can do to stop separating ourselves from the whole in which we live will inevitably lead to greater health. One of the most effective ways to use chanting in this regard is as a means of attuning our deepest feelings with whatever image we have of the one wholeness (holiness). The sound OM is often used because it is a representation of wholeness. Its components A-U-M can all be made by producing one sound, Ah, in the back of the throat and shaping the lips to change it into the variations. Try the following.

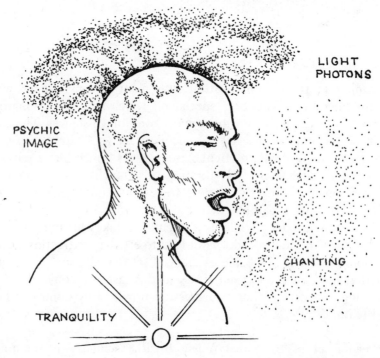

* "May You Live in Health" by Dr. Aaron Friedell gives medical instruction for scientific deep breathing. Available from University of the Trees Press.

First produce an "Ah" by letting the sound rise from the lungs below the larynx. This is very important because as soon as we start to produce the sound consciously with the larynx we are singing, and singing brings all the associations with "sounding good" into the act; this is not the purpose of chanting. Chanting is concerned with getting away from all the mental associations and expressing the deepest feeling sides of our nature which are normally shut up by the constant chatter of our mentations. The kind of "Ah" we want to make is the same kind we make for the doctor with a tongue depressor in our mouth. Use the Great Yogic Breath method outlined above and produce the "Ah" with the minimum amount of air so that it will continue for a long time.

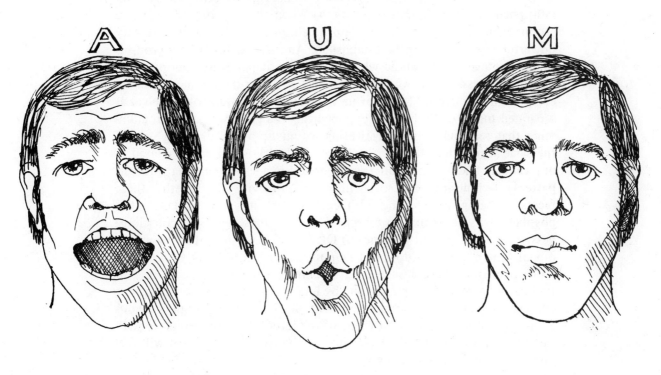

The next step is to shape the mouth and lips into the shape we use for making a "U" while still producing the "A" sound, with the relaxed throat, broad and flat at the back of the tongue. This rounding of the lips produces an "O" sound. Practice making this sound with a friend so that you can tell if you are successful. A tape recorder will also work. The final step is to shape the mouth and lips into the shape used for humming, lips together, while continuing to make the same "A" sound from below the larynx. This will produce an "M" sound.

There are several processes occurring while chanting which work together to produce profound changes within the chanter. 1) As the sound is carried on the long wind from the lungs there is a calming effect on our entire nervous system and enhanced receptiveness to cosmic energies. 2) As we chant we hold an image of what the whole is to us which both tunes our mental vibration to resonate with it and programs all the cells of our body to respond to it. 3) Whatever is expressed on the outbreath will be impregnating the surrounding air molecules so when they are inhaled, we will be getting right back what we sent out. The bonus is that each air molecule, like a holographic chip, contains a record of the whole so that the inbreath will be carrying the message we sent out back in, multiplied many thousands of times. 4) When the chanting is done as a group exercise the effects are multiplied many times over and very powerful resonances can be established. Another factor with a group is that the image of the whole which all the participants are concentrating on and expressing can work immediately to forge deep bonds between everyone in the group. Many of us walk around thinking of ourselves as separated lumps of flesh and group chanting is a great way of deprogramming that self-image and substituting an image of us all being part of the same whole, like fingers on a hand. 5) Health is intimately related to our cellular and organic functions, our self-images, our mental thought patterns, our desires, our social life, our sensitivity and whether these facets of being are in tune with each other and nature's laws. It is possible to use chanting to reprogram any of these levels of consciousness when they get out of tune. 6) Of course, if we do not know we are out of tune and sick it does not help us very much to know that we can reprogram our consciousness. But chanting can also be a means of finding out whether we *are* in tune by letting each level of consciousness express itself in turn during the chant. As chanting is a spontaneous expression of our being, if any one of the levels is out of resonance with the whole we will get a strong feeling of that in the chant -- it will feel out of tune and uncomforting.

The key to chanting is that it is a spontaneous communion between our little separated self and our big whole Self, an open expression of how we feel NOW in the core of our being. Naturally this is a difficult communication skill for many of us to master; in fact many of us are not even interested in taking personal responsibility for our health and would rather unload the responsibility on a doctor by paying out money for services.

Deep within every atom, cell and person is a fundamental vibrational pattern which results in what we see as a form. Chanting is a way of getting in touch with our fundamental vibrations and expressing them as they are, without any mental programs filtering and distorting them. The formal structures in which we have locked our awareness (cities, families, laws, words, expectations and so on) can be seen in this context as distortions of the true keynote within each individual. The terror, misery, pain and boredom that pervade our human social forms are all created by individuals. As our own bodies show us by illness, these vibrational patterns radiating from individuals are very often out of tune with fundamental notes. The theory of chanting is that the note is there and does not need any of our mental programs in order to be expressed, it is just that our minds have got us fooled into believing that the only way we can contact our true essence is through thinking.

Chanting OM is a way of bypassing all that mental garbage and letting the real note that is there ring out. After all, if all our mental acquisitions are doing us any good, why are so many of us running around searching for happiness and fulfillment? The answer is that the very source of all our problems is the one thing we trust the most and do not want to get rid of -- our mind and its self-image. While it will take much practice and persistence to get to the point where we can just express without having mental worry tapes about whether we sound good or are out of tune, eventually the real notes break through and we experience first-hand the peace of wholeness.

As the field of chanting is a profound study we suggest to those who are interested in the benefits of chanting that they listen to the series of tapes about the history and techniques of creative chanting available through the University of the Trees. At the University a choir has been training for three years in these methods under the leadership of Christopher Hills, who is reviving the style known in Sanskrit as Troupad (Words of Truth).

The promise of chanting lies in its vibrations, for this technique coupled with visualizations of our being linked up with the entire cosmos can help us make a quantum leap beyond our petty individual problems into a grand orchestra encompassing every note and sound in the universe. The bonus is that once we drop our identification as a separated individual lump of skin and bones beset with troubles, solutions to our problems usually flash into awareness and we see that in reality they are petty.

SEEING SOUND

It is helpful to keep in mind while studying vibration and form that vibrations can destroy just as easily as create forms. Take the familiar example of a high pitched voice breaking through a crystal goblet. Another famous case is the destruction of the Tacoma Bridge in Washington State due to the steadily increasing vibrations caused by a 45 mph. wind. The vibrations the wind set up in the suspension supports of the bridge were just the right rate to set up a resonance with the vibrational structure of the supports. The vibrations kept reinforcing each other until they got so strong they shook the bridge to pieces.*

Many unique forms are created when different vibrations are applied to mediums such as a body of water or a steel plate. One example is that of the tidal waves in Japan which are caused by huge chunks of rock falling into the Pacific Ocean off the coast of Chile in South America. Although the waves travel for thousands of miles, the water itself merely goes up and down. It is the vibrations caused by the tumbling mountain chunks that produce the waves and the motion of the waves.

A modern application of this relationship between vibration and form is the common phonograph pickup, which consists of a crystal attached to an electrical circuit. When the crystal's molecular structure is disturbed by dragging it across the record's surface it reacts by producing a minute electrical current which is channeled through an amplifier into a speaker which transforms it into a sound. In the jargon of electronics and physics this is referred to as the piezo-electric effect.

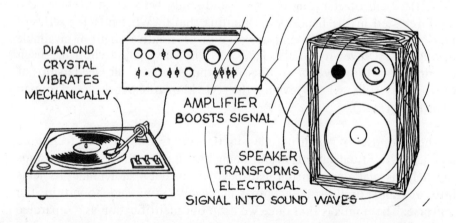

* Physicists discovered in their investigations into vibratory systems that when a system absorbs a vibration of the same frequency the oscillations reinforce each other and cause a "standing wave." The amazing quality of a standing wave is that because it keeps on reinforcing itself it can cause matter to either vibrate to pieces or rise to a higher energy state. Great singers take advantage of this by coupling phonetic cavities such as the mouth, nose, sinuses, and larynx in such a way that the sounds they are producing are amplified. The same principles are applied in the art of chanting.

MAKING A CHLADNI PLATE

A version of Chladni's original plate experiment can be put together quite easily with a plate of good bell material such as hard aluminum, brass, or steel. As can be seen from the diagram below the key factor is that the plate be firmly clamped in the center. This can be accomplished by attaching the plate to a threaded rod with a lock nut on either side. A bow of some kind, either a home-made bow or an old violin bow will also be needed to create a vibratory pattern in the metal plate.

Next, sprinkle a layer of fine sand on the plate and draw the bow across one edge. The sand particles will form into groups in the zones of least motion and create beautiful patterns. These patterns will vary with the different strokes of the bow. The edge of the plate should be grease free and the bow strings rubbed with rosin for best results. Touching the plate at different points while vibrating it will also result in changes in the formations in the sand. This is an easy way to see a vibrational system and interesting variations can be done by varying the shape and size of the plate into squares, rectangles, triangles, circles, ellipses and so on.

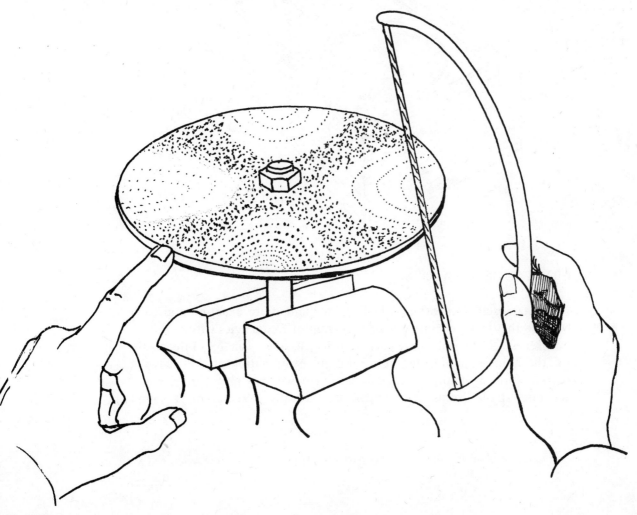

CREATE FORMS WITH YOUR STEREO

The cone of a loudspeaker will show similar effects to the Chladni plate by coating it with a thin layer of powdered sulfur or lycopodium. The ideal cone is supposed to vibrate as a whole but the paper cones commonly used in speakers rarely do. As the frequency is increased the movement is limited closer and closer to the center. Records with different frequencies recorded on them from 50 Hertz to 15,000 Hertz are easily available through any record store. Another way of generating different audio frequencies is to hook up a sound generator to the amplifier and out through the speaker(s). The advantage of a sound generator is that it allows infinite variations and fine tuning. The patterns on the cone will depend on the thickness and composition of the material as well as how they are suspended at the center and edges.

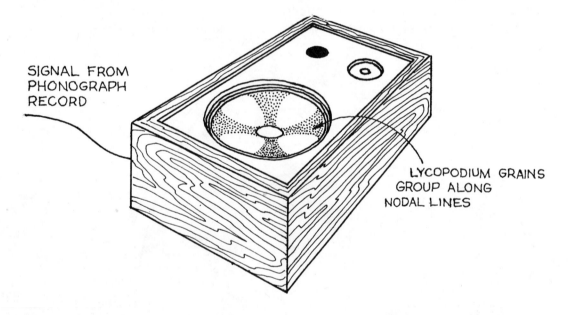

CRYSTAL OSCILLATORS

Crystal oscillators offer a much wider and more exacting approach because they can generate a wide spectrum of frequencies with excellent preciseness. Their only drawback is price, as they can be quite costly for the home experimenter. A continuous examination of the changing of one form to another can also be made whereas prerecorded tones generate stable forms. The apparatus basically consists of a sound generator coupled to the crystal oscillator, the oscillator in turn being attached to the underside of a steel plate.* The first step is to hook up the necessary equipment and cover the surface of the plate with a thin layer of sand. The next step is to start feeding different frequencies into the plate and watch what happens. The infinite variety of frequencies yields an endless flow of different forms.

Watching the sand flow "fluidly" as if it were some liquid current from one form to the next is fascinating. Equally intriguing is how the sand can flow in opposite directions at the same time. An easy way to see the currents flowing within a pattern is to mix colored particles with the sand before applying the vibrations. Forms and patterns which appear to be stable and stationary with unmarked sand can then be seen to be revolving, circulating and spiralling vortexes.

DEFYING GRAVITY

The sand or powder that we are using is naturally forced into the form corresponding to the vibratory stimulus being fed into the steel plate. While the tone persists, the substance (this works with liquids and viscous masses as well) remains in that form even if we hold the steel plate vertically! Be careful not to stop the tone while the plate is vertical

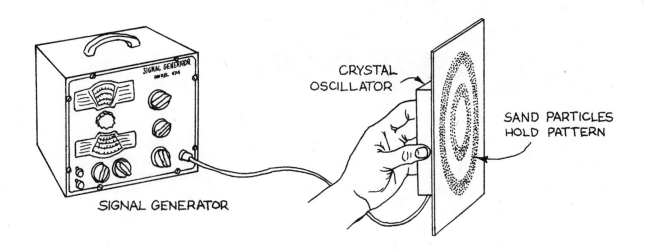

* Actually many other materials can also be used in this type of experiment. Particularly interesting mediums include thick viscous liquids such as glycerol or oil.

because all the sand will fall off the plate. If the disruption is short enough however, the sand in some instances will even "climb back." This is the equivalent to an anti-gravitation effect because the energy field of the vibrating plate exerts a stronger influence than the earth's gravitational field.

LIQUID FORMS

Whereas with solid mediums the line patterns correspond to the nodal lines of the wave fields produced by the vibrations, an entirely different result is obtained by using liquids. With liquids the antinodes appear as wave fields and the nodal lines disappear.

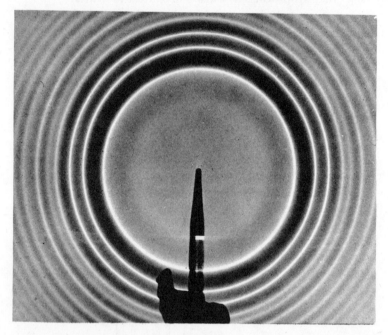

Circular ripples in a ripple tank are started by vibrating the tank.

Different viscosity liquids give a very wide range of forms and structures, viscous liquids in particular taking on the various wave patterns that reflect exactly the sound from the crystal. Liquids also have the advantage of being able to be immediately transformed by changing vibrations, and music can become a dynamic sculpture of movement as the vibrations of a symphony or sonata are impressed onto a film of oil or glycerol.

MOVING SCULPTURES

An exciting addition to these studies is the use of photography to record patterns. Highspeed photography enables forms to be seen which would ordinarily escape the eye and by using a stroboscope as a light source we can easily take high speed photographs. Motion photography is even more exciting because with it we can capture the moving sculptures being created by the vibrations.

Many effective variations to work with can be suggested. Both flame and smoke are easily modified by vibration. Sounds such as voices or music played onto a candle flame or onto rising smoke produce attractive and dynamic forms that reflect the sound itself. Linking an oscillator to a human voice can vividly illustrate patterns we create when speaking: a well pronounced 'O' for instance, produces the same shape in the sand on a Chladni type plate. Vibrational force fields and magnetic force fields can also be studied quite effectively by using iron filings because they often form into elaborate three-dimensional sculptures which outline the field effects.

SEEING SPEED AND WAVELENGTH

Another simple and beautiful experiment devised in 1876 by August Kundt allows us to see the speed and wavelength of sound waves. The apparatus needed for this test is quite simple to make. As can be seen in the diagram below the equipment consists of a long glass tube (a long fluorescent tube with its ends cut off and the phosphors inside it removed will do the job nicely), a piston at one end, a diaphragm at the other end and a long steel rod mounted so that it lightly touches the middle of the diaphragm.

In cleaning out the tube be careful not to inhale the contents, as they are highly toxic. After it is cleaned place a small quantity of lycopodium powder or precipitated silica inside the tube; it will indicate the force lines of the vibratory field. One end of the tube is stopped up with a moveable piston which will regulate the tube length (be sure it is a good fit). The other is covered with a thin rubber or cellophane diaphragm against which the steel rod is allowed to make a very light contact. The steel rod is clamped to a stand in its middle so that the other end can be rubbed to produce a sound.

Now stroke the outer end of the rod with a rosined chamois, making it vibrate. With a little practice it is easy to make the rod "sing" out a strong tone. This vibration will in turn set the diaphragm vibrating, causing the powder inside the tube to dance around. By slowly moving the piston at the other end of the tube the length of the vibrating column of air inside the tube can be tuned to resonate with the rod tone; that is, standing waves will be formed inside the tube. This living column of air is the active factor in instruments such as the saxophone, clarinet and flute and often one hears musicians describing it as "alive."

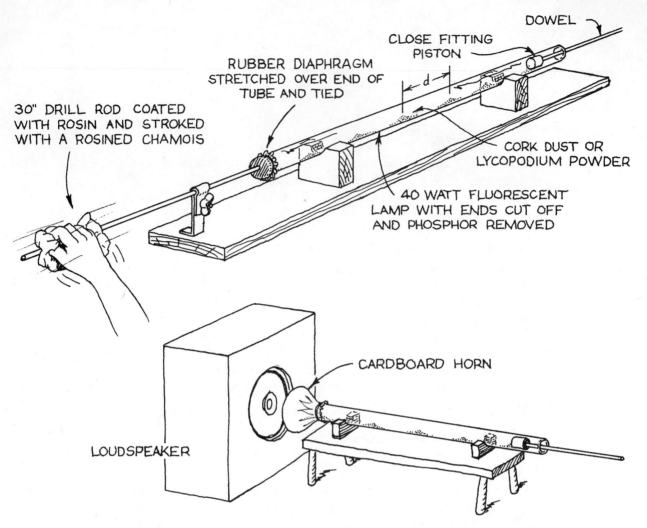

The resonance tuning of the piston will be quite noticeable as there is a dramatic increase in the volume of the tone. Another fascinating effect is that the flying powder inside the tube will settle and collect in little piles at the "nodal" points where the air is not vibrating. One node will be at the piston end of the tube, another at the diaphragm, and others in between, depending on the tone. The distance between these nodes is half the wavelength of the tone, therefore twice the distance is equal to the wavelength. As explained in Chapter Ten, it is also possible to discover which color band the tone resonates with.

To discover the speed of a particular tone it is necessary to know the frequency according to the formula: wavelength X frequency = velocity. This is easiest to do by substituting a cardboard horn for the steel rod (see diagram) and aiming a loudspeaker at the horn. The horn will cause the diaphragm to vibrate the air column. The advantage of this variation is that a record of known frequencies (available through any record shop) can be played into the horn. Because the frequencies are already known it becomes a simple matter of arithmetic to compute the velocity according to the formula given above. Additional variations include substituting specific gases for air and varying the length and composition of the rod touching the diaphragm.

ENERGY, VIBRATION AND FORM

The phenomena illustrated by these experiments involve three interrelated processes: 1) energy/power 2) wave/vibration and 3) form/configuration. While in the ending phase of the cycle patterned formations appear and in the beginning phase dynamic energy processes are generated, the whole cycle is initiated and maintained by the periodicity of the vibrations. These three aspects are all derived from the vibrational phenomena in which they appear and cannot be regarded one without the other. Invariably all three aspects are present, like three entwining strands of rope. Although one may predominate at any given moment, if any one is not present the other two cannot by definition exist, for forms vibrate, vibrations take form, and both arise from the energy in the Voice of Creation.

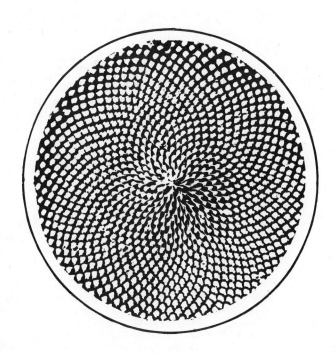

The transcendental logarithmic spiral of the Fibonacci Series, which is at work in the spiraling of a galaxy and the Pi-ray, is also in the simple sunflower.

RAYS OF LIFE

THE GREAT PYRAMID

On a flat, mile-wide plateau ten miles west of Cairo, Egypt stands the Great Pyramid of Cheops, a true wonder of the world. The base of the pyramid covers an area of thirteen acres all of which is leveled to within one half an inch. This famous pyramid was constructed from two and a half million (2,500,000) limestone and granite blocks weighing a total of six million tons. Each block was precisely cut for its own exact location by some yet unknown means. The joints were so finely cut that the space between them is no more than one thousandth part of an inch over an area of up to thirty-five square feet! These granite blocks are so hard that it takes a diamond drill bit under two tons of pressure to even make a dent in them. The enormity and precision of the great pyramid indicate that the ancient Egyptians were either more advanced technologically than has been assumed, or that they had help from an advanced culture, possibly from the legendary Atlantis. Because the only known device today that could carve such huge blocks so accurately is the laser, some students of the pyramid are paying closer attention to ancient legends that the Egyptians were visited by a people possessing supersense and an advanced technology.

The architecture of the Great Pyramid is intricately precise because it incorporates certain fundamental principles of cosmology and mathematics which were supposedly not known to mankind when it was constructed around 4,200 years ago. The circuit of the base, for instance, measures 36,524 pyramid inches.* This figure is the length of one

---

* The high central section of the King's Chamber Passage (known as the antechamber) consists of a square measuring 5 x 5 Royal cubits. Each Royal cubit equals 20.606593 Pyramid inches. A circle of the same area as this square measures 365.242 Pyramid inches around its circumference (i.e. 1 pyramid inch for each day of the year.) The Royal cubit equals 20.6285 British inches. Therefore 1 Pyramid inch equals 1.001064 British inches. The British inch presumably derives from the ancient Pyramid inch but has changed very slightly over the centuries. The other measurement used in the designing of the Pyramid is known as the Sacred cubit which equals $\frac{10\sqrt{\pi}}{4y}$ Royal cubits. The Sacred cubit is one 10,000,000th of the mean distance from the center of the earth to the poles. Observation by satellites measure the distance as 3949.9 miles and this divided by 10 million gives a result of 25.0266 British inches, correct to four decimal places. Hence the Pyramid inch is one 500,000,000th of the Earth's polar diameter. The more recent attempt to base the unit measurement on the earth size was the introduction of the meter as 1:10,000,000th part of the quadrant from the North Pole to the Equator. However as the earth is not spherical the meter is not as accurate as was hoped. The Pyramid inch is based on the only constant straight line in the earth -- the axis.

terrestrial year or 365.24 days. The Egyptian calendar included an extra day every four years to make up for the extra one quarter of a day each year, just as we do with our leap year. The sum of the base diagonals equals 25,827 pyramid inches. The precession of the equinox has been calculated at 25,827 years. The Great Pyramid was probably used as a major astronomical observatory which was not equalled in accuracy until the invention of the telescope in the seventeenth century. In fact, Soviet archeologists are reported to have found perfectly spherical crystal lenses of great precision which were possibly used in telescopes in ancient Egypt, though with their methods of divining they would not need optical telescopes to gain accuracy.

Geometrically the structure of the pyramid incorporates the value of the endless ratio, pi (symbolized as π and equalling 3.1416 . . .) which is the proportion of a circle's circumference to its radius. The pyramid's height is proportional to the perimeter of its square base in precisely the same ratio as pi. By circling the square of the pyramid's base, the resulting circle has a radius identical to the height of the pyramid. In other words, the pyramid successfully squares the circle and triangulates the half-sphere whose central point corresponds with the apex of the pyramid.

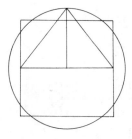

The square symbolizes the 'earth' of matter and rationalism. The circle symbolizes the encompassing world of spirit, heart and feeling. Thus the squaring of the circle was the architectural as well as philosophical pursuit of the ancient sciences of religion.

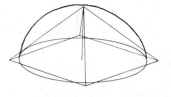

The Great Pyramid includes the 3,4,5 triangles which made Pythagoras famous. In addition, the structure is also based upon the fundamental proportion of the "Golden Mean" or sacred cut by the Greek letter phi (symbolized as φ and equalling 1.618 . . .). This formula is the basis of aesthetic proportion used by artists and architects down through history to the present.

Phi is obtained by dividing a line AB at point C so that the whole line (AB) is longer than the first part (AC) in the same proportion as the first part (AC) is longer than the remainder (CB).

   AB is to AC as AC is to CB

Phi is an endless ratio, just like the mysterious pi.* Plato in his *Timaeus* considered this the most binding of all mathematical relations, and called it the key to the physics of the cosmos. The phi proportion occurs throughout nature. Our body is divided in this ratio. Leaf distribution in plants follows this ratio and is known as the Fibonacci series.

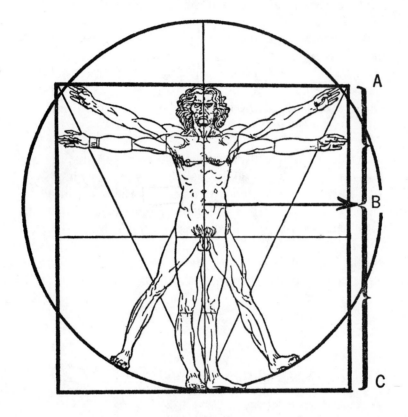

The famous drawing by Leonardo Da Vinci vividly illustrates the correspondence between the sacred cut and our body.

Most important of all from the standpoint of sacred, magical geometry, the diagonals of the five-pointed star divide each other in this ratio, which can be seen extending infinitely into both the micro and the macro.

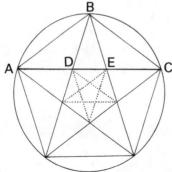

AC/AB = phi
AD/DE = phi
CE/DE = phi
etc.

* Both pi and phi can only be expressed as imperfect functions, as relationships. They are directly related mathematically based upon the formula $\pi = \varphi^2 \times 6/5$ (2.618 x 6/5 = 3.1416...).

The mythical five-pointed star is a national emblem which appears in the flags of the three world superpowers: the United States, the Soviet Union and the People's Republic of China. The pentagram, when drawn on the ground with its base to the north and its chief point to the south is useful for black magic purposes for it generates a negative energy spiral. Conversely, with base to the south and pointed toward the north it generates a positive energy spiral useful for beneficial, white magic purposes.

PYRAMID ENERGY

The basic secret of the great pyramid is that *it is a five-pointed star in three dimensions.* Five-sided and five-pointed, it stands with its apex pointed to the sky, to the beyond. It is at this apex or fifth point that powerful vibratory energy fields are focused by the pyramid form acting as a psychotronic lens. Pyramid research using the methods of radiesthesia and supersensonics with scale models of the Great Pyramid orientated to magnetic north has determined that there are two basic energy spirals originating at the area of the pyramid's apex where the capstone was located. A positive, clockwise rotating energy spiral comes off the top, while a negative, counter clockwise rotating energy spiral radiates down inside to the base.

Early radiesthesists Chaumery and Belizal found that a half-sphere had similar mysterious properties as a radiator of energy when placed on a flat surface with the curved side uppermost. They found a positive energy spiral flowing upward from the top that corresponds with *prana, qi (or chi)* or *life energy.* The same energy can be detected radiating from healthy plants and animals. This positive life energy was found to lie in the visible green part of the vibratory spectrum of color harmonics and for this reason it was called "positive green." At the same time, they also noticed a downward flowing, counter clockwise negative energy spiral of

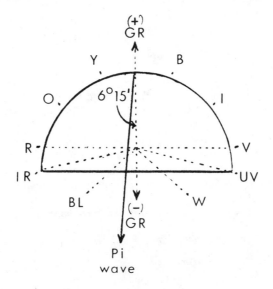

disintegration which they detected emanating from the half-sphere at a point diametrically opposed to positive green and therefore called it "negative green." By testing various objects used by ancient civilizations for religious purposes, they found that many of them emitted strong radiations in these green wavefields. This was especially true of scale models of the Great Pyramid orientated to magnetic north, which we have seen reproduces, in flat surfaces and angles, the curved contours of a half sphere. This discovery is of profound importance for thousands of people who are presently meditating inside and sleeping under pyramids thanks to the premature and unwise solicitations of pyramid manufacturers.

Much interest has been shown in using the pyramid energy as an aid to meditation, mostly by people who know nothing about meditation or radiesthesia and are better at promoting commercial sales than research. The fact is that beginners when sitting under the apex of a pyramid do "feel" something happening inside their head and in their spine... All these effects are dangerous, not only from the psychological viewpoint which is that a self-suggestion of increased spirituality by these means is of negative value and merely feeds self-righteousness, but they are also psychophysically harmful, because of the downward rays of negative green energy.

Christopher Hills, "Supersensonics"

Negative green is the cause of most sickness and mental disturbance because it resonates with and increases the downward flow of psychic electricity and brings dehydration, cessation of cell life (even cancer) and what is called mummification. This is the direct opposite to the positive green upward flow of prana to the magnetic part of the mind which brings integration and unity of body, mind and spirit. The downward flow is known in Yoga by the Sanskrit word "Apana." This is the psychic centrifugal force which moves through every system of material particles causing disintegration and decay, while the practice of reversing this flow, pratyahara, integrates, synthesizes and prolongs the complete unity of the system. In meditation negative green results in restlessness, lack of concentration, abnormal sexual appetites, etc. On the other hand, the positive green life energy is flowing through us at all times and we do not need to sit on top of a pyramid to tune in to it and benefit from it.* Negative green has constructive purposes, however. There is within the negative spiral a narrow band which is both life-giving and life-taking —the pi-ray.

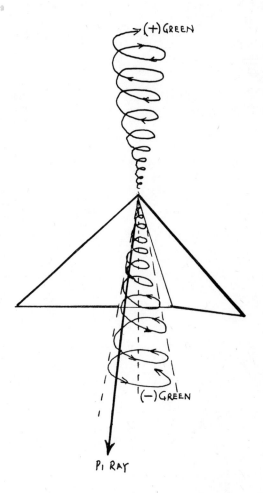

THE RAY OF LIFE AND DEATH

The Great Pyramid has a sarcophagus of granite in the royal chamber which has four enormous stones (estimated to weigh as much as 600 tons each) placed over it with the vibration of a half sphere as a roof on top of them. This special chamber is set off from the central axis of the pyramid about 10 meters to the south side at an angle of 6° 15' off the east-west plane. This structure was specially designed to block out most of the negative green, and acted as a resonating chamber allowing a special ray, the pi-ray, to fall on the sarcophagus, or coffer.

The word pyramid comes from the Egyptian hieroglyph PR-MS which means to "arise at birth" and thus the pyramid is a form or pattern of energy which symbolizes rebirth or resurrection. The coffer, irradiated by the pi-ray, was used to test the spiritual condition of initiates to the Egyptian priesthood.

---

* Experiments are included at the end of this chapter so that you can prove these points for yourself.

After many years of training and spiritual tests, the ray of life and death was used to test the initiate's detachment from worldly vibrations ... The length of time one could withstand the effects of Telluric rays was an indication of purity of consciousness. The religious symbol for this ceremony is the weighing of the heart on a balance with a feather on the other end.

Christopher Hills,* "Supersensonics"

Bettmann Archive
Portion of Egyptian Book of the Dead. *To the left a soul is being weighed, one of the tests to be passed before the soul entered into fellowship with Osiris and the other gods*

THE AMAZING COFFER

As we saw in Chapter Five, Christopher Hills has invented a pyramid energy box called the "Pi-ray Orgone Energy Accumulator Coffer" which gives truly miraculous powers to its operator. The design on the top of the coffer when seen from above represents the shape of the pyramid. This symbol produces the same energy as the pyramid form so a physical pyramid is unnecessary. A positive copper spiral lining the inside cancels the effect of negative green and allows only the pi-ray to descend into the box, thereby isolating the concentrated life force.

The coffer is activated by the consciousness of its operator and will amplify whatever is present in that person's mind whether positive or negative. Those who purchase the coffer are warned to be sure they have a positive attitude, for whatever we activate in our consciousness will be amplified in the result. In addition, the coffer is specially designed so that it can only be used for good. Anyone who deliberately tries to use it for a harmful purpose will experience an immediate return shock. The coffer must be checked periodically with a pendulum reading to insure that the powerful pi-ray is not used to excess, for once the saturation point has been reached what was previously curative or beneficial becomes harmful. If we use it to charge ourself up, for instance, we must be sure we do not get too much energy and become manic or unable to get to sleep at night.

* Christopher Hills spent a night in the King's Chamber in 1960, where much of this previously secret knowledge was revealed to him.

 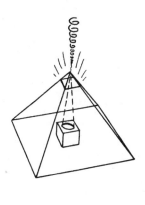 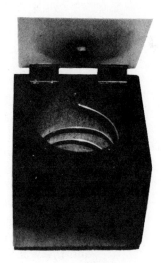

The coffer is a broadcaster of vibratory energy patterns which works on the principle of the "witness."

> The word "witness" does not mean a person in a legal court case but an object which is an exact replica or duplicate of any sample or situation. A piece of hair is a witness of your body. A blood spot is a witness of your body. A photograph is a witness of the object photographed.
>
> Christopher Hills, "Rays from the Capstone"

By placing an appropriate witness in the coffer, aligning it with magnetic north and placing it where the direct rays of the sun can be focused by its mirrored capstone directly down the hole in the center of the lid, it can begin to perform some truly amazing feats. Like the legendary Aladdin's lamp, it contains a genie -- the pi-ray -- which you can harness to transform your life.*

A NEW ORDER FOR A NEW AGE

Many of the founders of the United States of America were Masons, whose Hermetic tradition goes back to ancient Egypt and the building of the Great Pyramid. In choosing a seal for the country they decided upon the symbol of the pyramid with a capstone floating above it and with a radiant eye looking out of it. Why did they choose this particular symbol?

---

* Many successful experiments with the coffer have been conducted by the student/faculty of the University of the Trees and people who have purchased it. These experiments prove the ability of the coffer to operate over a broad range from healing to altering consciousness to buying and selling property. Many of these experiments along with a more detailed explanation of the coffer and what it can do are covered in "Rays from the Capstone."

The United States Treasury says that the eye depicted in the capstone of the pyramid represents the eternal eye of God, but this interpretation is incomplete. Actually it is a symbol for what lies behind all living eyes and all loving action. It represents that radiating light of consciousness which has become single and ONE. As Christ said: "Make thine eye single and thy whole body will be filled with Light." He was referring to the light of consciousness, symbolized in the Great Seal of the United States of America, which also contains Latin inscriptions which point the way to our destiny.

The inscription over the capstone is *Annuit Coeptis,* which means "God favors our undertaking." Americans must ask -- what is our undertaking? This is revealed by the inscription written under the pyramid: *Novus Ordo Seclorum* - "A new order of the ages." America is the focal point for the New Age that is dawning in more and more of us. We carry on that revolutionary spirit of individual freedom of our forefathers which is the very foundation of America. It is this spirit of freedom and revolution living in each of us that is the real America, not the material trappings of flags, warships or armies, nor the vain power trips of our "leaders." It is up to us to be our own leader, to find peace and freedom within ourselves, if we are to give freedom to others and to the world. Only when we realize that it is the light of consciousness streaming out of our eyes that is the real key to our conscious evolution, will we find our way back to the ONE, that divine, central source of all peace and all freedom.

EXPERIMENT SECTION FOR CHAPTER EIGHT

EXCESSIVE CLAIMS

Many claims are being made today about the miraculous benefits of pyramid energy (usually by people interested in profits rather than serious research). "Ants and other insects will not bother food stored in a pyramid." From experimental evidence, this claim appears to be true. However, few people realize why insects stay away from pyramids. Insects are not as dumb as we think. After all, would you enter the area of a death ray which saps your life energy?

CLAIM NO. 1: "Food does not decay inside a pyramid." This claim also appears to be true. It results from the mummifying power of negative green which dehydrates and kills the bacteria which cause decay. This effect was discovered in the 1930's by a French radiesthesist, André Bovis. While visiting the King's Chamber in the Great Pyramid, he noticed a trash can containing the bodies of animals which had wandered into the pyramid, lost their way and died. He noted that these bodies did not have the characteristic putrid smell of decaying flesh and that they had become mummified and were remarkably well preserved. He built a scale model of the Great Pyramid, orientated it to magnetic north and placed a dead cat in it. Over a period of several weeks he found that the body of the cat did not decay, but rather became dehydrated and mummified. You can test this effect of pyramid energy yourself in the experiments which follow later.

CLAIM NO. 2: "Sitting inside a pyramid improves meditation and expands consciousness!" A good test of this claim is to see if there has been any real positive improvement in peace of mind and direct heart contact with others.

CLAIM NO. 3: "Sleeping under a pyramid cures insomnia and promotes astral travel!" This may be true, thanks to the energy draining effects of negative green, but does it also bring about erotic dreams and wild fantasies?

CLAIM NO. 4: "You can actually *feel* the energy!" True, many people do report experiencing something inside a pyramid, usually an intense heat. This is the dehydrating, disintegrating negative green at work. The powers of autosuggestion and self-hypnosis are very great and have led many a student on a downward path thinking that something other than his or her own expectations were at work. The claims of increased spirituality and well-being supposedly resulting from pyramid meditating and sleeping need to be checked against the real changes in personal being that have resulted. Insects such as ants seem to be more intelligent than people when it comes to negative green energy because they stay away from it. The effects of negative green can be detrimental to your health. In supersensonics negative green is found to resonate with and indicate the presence of disease, especially *cancer*. The best way to find out the veracity of these claims is to build a pyramid and find out the truth for yourself.

BUILDING A PYRAMID

The best pyramid form for experimental purposes is a scale model of the Great Pyramid. Pliable cardboard is best for small models and plywood is fine for larger ones. The following chart gives proportional dimensions for several different sized pyramids:

BASE	SIDE	HEIGHT
6"	5.7"	3.8"
12"	11.4"	7.6"
24"	22.8"	15.2"
36"	34.2"	22.8"
72"	68.5"	45.6"

The formula for determining the size of any pyramid in the correct proportions is: length of base side minus five percent equals the length of a slanting side.* Be sure to make a base as well as the four sides, and leave one side open for putting in and taking out samples.

After building your pyramid (the 6" size is sufficient for these experiments) make an orientation map by tracing the square base on a larger sheet of paper or cardboard. Draw two lines -- one north-south and one east-west -- which cross exactly in the center of the square and extend out beyond the base so that you can see the lines after you position your pyramid on the square. The area where the lines cross lies directly under the apex. You may also wish to draw a similar cross on the bottom inside of your pyramid in order to know exactly where to place the material you will be testing. Orientate the base of your pyramid with magnetic north.

MUMMIFYING FOOD

Choose two identical pieces of meat -- hamburger will do. Make sure they are the same size and both weigh the same amount. Place one piece

---

* "Handbook of Psychic Discoveries," Ostrander and Schroeder.

inside the pyramid in the center.* Place the other one to the side at least several feet away. This will act as a control, against which to compare your results. Allow from two to six weeks for this experiment. At the end of an appropriate period of time notice the condition of the two portions. Weigh each and note any difference between them and their previous weights. Taste the meat that was kept in the pyramid (but not the control), if you wish. Some researchers claim that meat that has been mummified in a pyramid remains edible for quite some time. However, the nutritional value of such meat is questionable for negative green saps the life energy of food as well as people.

You may also wish to compare pyramid-dried fruit with sundried fruit to see if there is any difference in taste, nutritional value, etc. Simply repeat the above experiment (keeping the pyramid out of direct sunlight) substituting fruit for meat and place the control sample in a sunlit spot.

SHARPENING RAZOR BLADES

Karl Drbal, a Czech engineer, has patented the Pyramid Razor-Blade Sharpener which is now selling throughout Europe, amid claims that you can shave for months with a blade kept in a pyramid between shaves. Apparently, the negative green energy, by dehydrating the residue on the blade, slows the oxidation reduction processes that dull it. You can test this yourself by simply keeping a razor blade inside a small pyramid between shaves. This may be one of the few really beneficial uses for negative green since there is no problem with sapping the life energy in a metal alloy.

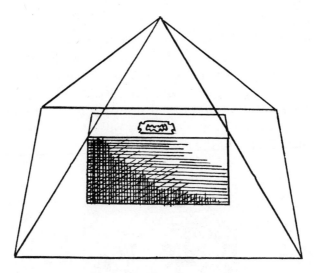

* Some researchers recommend supporting the item on a platform one-third of the way up (the approximate proportional location of the King's Chamber). But this is not necessary, for negative green is a cone-shaped spiral that extends all the way from the apex to the base down the center.

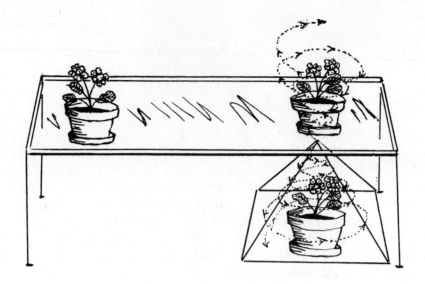

TEST YOUR PLANTS

Plants are extremely sensitive to vibrations. Therefore, a good way to test the effects of positive and negative green energy is to use plants. For this experiment you will need three small plants such as coleuses of nearly identical size and state of health. Place one inside the center of your pyramid. Place another on a platform on top of the apex or hang it directly above it. Place the third some distance away. Insure that all three receive the same amounts of sunlight, water, fertilizer and affection. Clear your mind of expectations for they generate mental and emotional vibrations which your plants will pick up and react to, thus interfering with your results. Take comparative notes on the color, posture, growth and leaf development each day and see for yourself how the pyramid affects the plants and their life energy as reflected in their general vitality. Watch the plant at the top of the pyramid closely. An overcharge of positive green may occur if the plant is left there too long.

A variation on this procedure is to place all three plants in the same location some distance away from the pyramid. Take a leaf from each one and label it as well as the plants so you know which leaf belongs to which plant. Place one leaf inside the center of the pyramid, one on top and the third off to one side at least several feet away. Besides noting the effects of the pyramid on drying and shrinking of the leaves, check the condition of the three plants. The leaves will act as witnesses to broadcast the pyramid energy to the plants.

HILLS' POSITIVE GREEN PENDULUM

A CURE FOR NEGATIVE GREEN

The Hills' Positive Green Pendulum detects all vertical waves, both positive and negative, and is excellent for a host of divining purposes. It consists of a round wooden ball which is painted with the color of positive green. Its long polyrod point acts as an intensifying wave-guide for energy emitted from radioactive powder (not harmful) contained in its center. It is specially designed and tuned to the same positive green vibration which comes from plants and the tops of pyramids, the prana of Yoga and Qi of the Chinese. With this pendulum you can learn when positive or negative green is emanating from something and in what amount. For example, most plants radiate positive green until the sun goes over the horizon. Healthy things emit positive green; decaying and cancerous things emit negative green. Both can be detected with the positive green pendulum.

This fact was vividly illustrated one day at the University of the Trees. The mother of one of the students was visiting the center and happened to pick up a positive green pendulum. She held it over her hand and it immediately began spinning in a negative, counter clockwise direction with a wide arc indicative of high negativity and possibly cancer. Several students then double checked this reading and got the same pendulum response each time. She was asked about her negativity and whether she was suffering from a disease. She answered that she was feeling fine and could find no reason for the pendulum response, which she characterized as rather "funny." A few weeks later, however, she called her daughter to report that a medical examination revealed that she had cancer of the cervix. She later underwent a hysterectomy.

This operation may not have been necessary if she had used the positive green pendulum to detect the negative energy before the cancer had developed. However, it is not too late for those who have been absorbing negative green by sitting and/or sleeping under a pyramid. All that is required is to learn how to tune the pendulum by reading the instructions in Chapter Ten and obtaining a positive green pendulum from the publishers. Once this has been done the procedure is rather simple. Merely rotate the pendulum over the sore or diseased area in a negative, counter clockwise direction. At the same time place the hand that is not holding the pendulum in a vessel of water. Set up in your consciousness the image of the negative green being channeled from your body through your hand and into the water. When the pendulum stops rotating the negative green is gone from your body into the water. Merely throw the water away somewhere safe. Now to charge yourself up with positive energy, or prana, simply rotate the pendulum in a positive, clockwise direction over your left (receptive) hand, breathe deeply and set up in your consciousness the image of the energy flowing throughout your body. When the pendulum stops rotating you are fully charged with energy. Do not force the pendulum to rotate after this for you will charge yourself with too much energy.*

* This Self-programming of positive green or prana should be repeated until the pendulum no longer rotates to the left spiral.

CREATIVE IMAGING

What did we see when we were fresh born babies? Did we see what we see now, all the sharply defined objects with their specific colors, contours and backgrounds? Did we see the world as "out there" beyond the limits of our skin?

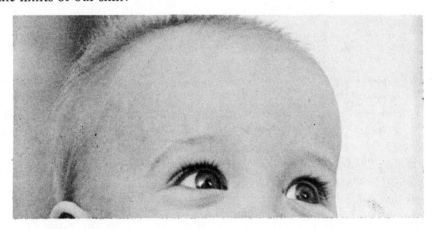

Somewhere in our highly impressionable youths we all went shopping at our local "belief systems stores" and bought ourselves an "ideal eye" which we have treasured ever since as a trusty guide to reality. As we have seen from the work of Ronchi and Hills discussed in Chapter Five our trust has been betrayed, but by whom? The usual reply is to lay the responsibility for our world models on the societal contexts in which we were raised, but this is merely a continuation of the problem because it reinforces the subject/object philosophy which separates us from reality. A new path is being broken by the trailblazing scientists of consciousness who see their own beings as laboratories of reality and it leads to that mysterious faculty known as the imagination.

THE TREE OF KNOWLEDGE

In our imaginations each of us is the creator of an elaborate mental representation of reality by which we order our behavior. The actuality is a constantly changing interplay of vibratory wave fields which are all expressions of the same source: Pure Consciousness. We tend to attach a special importance to a narrow range of these wave fields which we call matter and filter the rest out of our awareness.

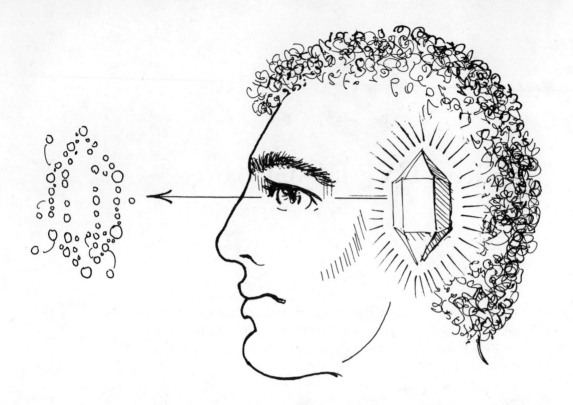

Most of us identify with the world of sensory stimulation which we experience through the limbs and branches of our nervous system in order to taste the fruits of the Tree of Knowledge of Good and Evil. But to eat the apples from this tree and not get hooked on them is very rare because their taste is so intense that we want more. How many of us are aware of the apple as a unique expression of invisible, untouchable light, crystallized via photosynthesis into a denser vibrational field? And how many of us are aware that the patterns we see, touch, taste, hear and smell are all images we are creating and experiencing within our consciousness?

REAL IMAGES

Our experience of the universe is a series of images in our consciousness which we have to share with each other in order to have a society. For this purpose we have created a set of symbols called words we use to represent our images. By tracing words back to their original meanings we can get a glimpse of the first-hand images of their authors which are often radically different from the current definitions. In this way we can make language a dynamic tool for investigating our own first-hand images of the world.

The roots of the word *image* offer us some good insights into the nature of images. *Image* comes to us from the Latin *imagio* which comes from the Greek *mayos* which is related to the Sanskrit *maya*. *Maya* is the illusory picture of reality which we derive from our physical sense percepts. *Magic, magi* and *mage* also come to us from the same roots.

There are two types of "image", *real* and *virtual*. A *real image* is the invisible primary radiation which contains a message while a *virtual image* is the psychic reflection of it which we can see. The shapes and forms that we see in our mind's eye are virtual images, reflections of the invisible primary radiations which we are continually interacting with but do not see. Everything in nature is a living energy system which is a unique transformation of pure light into a particular wave field which radiates its own special vibratory "signature." No two signatures are the same, yet all arise from the same source. This unique radiation is a system's real image and the continual crystallization of light which results in the tissues of our cells proceeds according to the real image encoded in each DNA molecule.

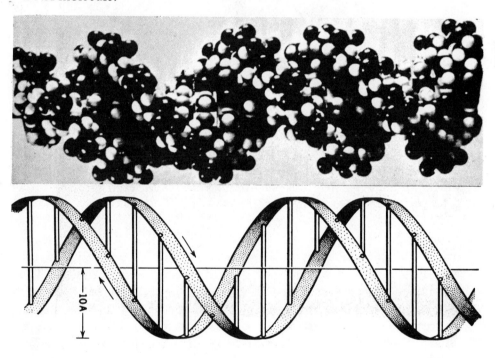

For us to see a true reflection of a real image is not easy, however, because the mirror of the mind usually acts like a stormy sea and distorts the image. This is why the virtual images that we see in our mind's eye are not to be taken as the real image and, hence, why it becomes important to know whether our "self-image" is a true reflection. The incongruency of our self-image with our real image can be seen in how often we either over-rate or under-rate our abilities, how one moment we are on top of the world and the next in deep despair. With a distorted self-image we are bound to create distorted virtual images in our consciousness.

In order to see a clear reflection we have to learn how to still the mind so completely that it becomes like a smooth pool of water, untroubled by the usual conflicts which rage within us. Most of these conflicts can be traced back to the problem of identifying our self as a physical body constantly bombarded by sensory stimuli. We often see ourself as a lump of flesh separate from other lumps of flesh, yet long to somehow bridge the separation and have deep communion with another.

One problem in accepting such a self-image is that because the physical senses are very limited in the range of vibrational fields which they can detect, we immediately cut off our awareness of most other life forms. Another problem is that a separated self-image leads to virtual images in which we are separate from the total environment, even though it is a scientific fact that we are inseparable.

Imbedded deep within our beings is the certain knowledge that we are united with the Whole so when we buy a separating self-image that goes so much against our grain we create a very serious conflict. The techniques of Creative Conflict then become valuable as tools for learning to see clearly, because they acknowledge this fundamental source of our physical, mental, emotional, and spiritual distress and help us get in touch with our real self-images. By bringing many different viewpoints to bear on a group member's conflict the "group mirror" penetrates the distorted virtual self-image and reflects the gap between it and the real self-image.

WHOLISTIC HEALTH

When the virtual self-image is a true reflection of the real self-image we are in a state of health. *Heal* is an old English word meaning to make whole, but the present-day usage belies it. At the core of modern institutional medicine is the allopathic theory of disease which holds that our suffering is caused by others, whether they be germs or people. The only way this view of healing can be believed is if we identify our self as a body separate from other bodies. Yet we know that if we do we create an inner conflict which in turn makes us unhealthy, not whole. The so-called revolution in wholistic healing is a swing in the opposite direction and increasing numbers of physicians and laypeople are seeking out new technologies and belief systems which support the model of disease as a personal responsibility. If we are to embrace a philosophy of health the first step is to acknowledge that any dis-ease is a reflection of an incongruency between who we think we are and who we are, and that the primary responsibility lies in our hands, not a doctor's. The virtual self-image and the real self-image have to be in balance for wholeness to be realized.

This principle of balance is the original truth underlying all of medicine and the ancient healers understood this and made it the core of their labors. If we trace the words *doctor of medicine* back to their original root meanings we can see this quite clearly. *Doctor* comes to us from an old Latin word meaning teacher, while *medicine* comes to us from two Latin roots meaning balance and ingested substance, in other words a substance for restoring balance. Putting these three together we find that a doctor of medicine is actually a teacher of how to achieve balance. Tracing the Latin root *medi* back even further through Greek and Indo-European we find the Sanskrit root *medha* meaning wisdom. To the great yogis of ancient India and to the pioneering medical sages of the Mediterranean cultures alike, the essence of wisdom lay in the principle and practice of balance in meditative action which led to a state of perfect health.

VIRTUAL AND SELF-IMAGE IN CONFLICT   VIRTUAL AND SELF-IMAGE UNITED

What is the secret of perfecting balance? The secret is balancing our virtual self-image with our real self-image by confronting the gaps in our self-awareness, the blindnesses and habitual thinking patterns which separate us from the total environment, and then to change these gaps.

MAGIC

A real magician is a person who understands that bridging the gap between our virtual self-image and our real self-image is the key to health and sets out to make the necessary changes. A key factor in the practice of magic* is the storing up and concentration of psychic electricity so that it can be focused in the process of change. Even more important is the will to change. Unless we really want to change, all the techniques in the world and the best teachers will be useless. Until we are completely fed up with our present situation and willing to do whatever it takes to conquer the habitual ways of thinking and relating which are causing us pain, it is useless to learn techniques. This is the problem with many of us today, we just do not want to change; we would rather suffer the pain of emotional, mental, physical and spiritual separation than give up our habitual compensations which provide us with temporary satisfactions or thrills.

An analogy is the dilemma which is presented to monkeys in order to trap them. A hollow coconut shell filled with nuts is tied to the branches of a tree and in it is a hole through which the monkey can barely fit his hand. The monkey comes along, finds the shell and eagerly sticks his hand in to grab the nuts. To his dismay he finds that when he gets hold of the nuts he cannot get his hand out because now it is too wide. He is faced with a simple choice, find food someplace else or hang on to the nuts in hopes that they will eventually come out. If he takes the first choice he surrenders a desire for these particular nuts, which is only causing him frustration, and moves on a free monkey. If he takes the second choice he hangs onto a desire that is not only frustrating him but also tying him to the tree.

If we liken the nuts in the coconut to the things we desire and the tree it is fastened to as the life situation we accept in exchange for gratification of our desires, how many of us will be like the free monkey and how many like the trapped one?

* See "White Magic: Spiritual Mastery of the Powers of Consciousness" by Christopher Hills, edited by Robert Massy.

Trapped by Desires   Free and Unattached

The magician is not caught in this problem because he is only interested in mastering the faculty which is creating all those external conditions, the imagination. The secret to successful magic is skillful action. Skillful action is taking our mind out and wrapping it around a project from the inside out, getting totally involved in what we are about to do in such a way that we actually imagine it happening and can see all the interrelations which are involved. One simple method for improving our skill involves the seven levels of consciousness.

In order to manifest a project on the physical level of consciousness we have to take the ultra-high frequency vibrations of the original image and step them down through all the chakras to the physical center. Most people make these transformations subconsciously and the end results can show which levels of awareness were operating and which were not. By consciously taking the image down through the different levels, however, we are able to make sure that all levels of awareness are involved in the process and the end result will reflect a much higher quality manifestation.

This type of work requires a single-pointedness of mind if we are to be successful. And conserving our psychic electricity instead of throwing it away into activities which do not contribute to our imagined goal helps considerably. Practically, this means that we focus the positive green heart energy on the image we want to manifest instead of letting it leak into habitual desires such as food, lust and ego-strutting. For example, if instead of venting our sexual lust everytime it builds up, we save the energy in our psychic reservoirs, we can bring our projects out of the clouds and down to earth with plenty of *roomph* to see them through to completion. If they do not come off after we have put our whole being into them, we know that somehow they were not in tune with the cosmos and we can move on to the next ones a little wiser. The end result of these magical practices is to refine our consciousness to the point where we become finely tuned instruments of the whole, reflecting reality instead of distortion.

HOW DO WE KNOW?

The study of the creative imagination is a big job requiring intense concentration and self-confrontation that many are not willing to undertake because it means first acknowledging that our favorite pastimes often serve to brainwash us and then making the effort to discard them. Self-confrontation also means questioning the origins of each image presented to our awareness in order to see what preconceptions are behind it. Once these concepts are out in the open, we question their validity by asking, how do we know what we know? Eventually this practice leads to questions about the origin of the one inside who thinks he or she knows something. Where does that inner "I" come from? Because the ways of the mind act like a labyrinth to confuse and turn the searcher of truth around in circles, the advice and guidance of one who has travelled the route already are invaluable.

BUDDHA   EINSTEIN

While we always have to do the work ourselves, the sage offers a living example for us to steer by and, because he knows where the reefs and dangerous shoals are hidden, we can avoid serious mistakes. One such reef is the ability of the mind's eye to see virtual images such as delusions and hallucinations that have no reference to the real vibrational fields we are sensing through our skin. The image we are dealing with right now in our minds could be an illusion with no counterpart in the world, so how do we know if what we are seeing is real?

This is a very deep question because it takes us into the realms of philosophy, the study of how we know what we know. The commonly accepted method of verifying our perceptions is social consensus; because a belief is generally accepted by many other people we buy it as an accurate representation of the truth. This method is not ultimately reliable however, because the herd instinct can often lead to a social consensus which is the exact opposite of truth, as can be seen in the examples of Nazi Germany, the Grand Inquisition in Spain and slavery.

TIME IS NOT REAL

Another excellent example of a false consensus belief is the notion of "time." Our entire social fabric is woven around time. Prompt and efficient organization of services, materials, people and plans all depend on the clock. Without clock time, trying to get various processes all synchronized so that things get done quickly and efficiently is impossible. If we relied on everyone's inner time sense for synchronization of efforts our society would fall to pieces within a month. "Time" is a mental concept created in human minds, an external standard to which we humans have agreed to conform. Time on our clocks is based on the cyclical rotations of the earth on its axis and its orbit around the sun. By dividing the intervals between these cycles into fifths, fourths, thirds and halves we create average units such as months, days, hours, minutes and seconds. These are conventions which we have adopted for the sake of simplicity, but they are not an accurate reflection of the actual cycles. When we divide the time it took for the earth to orbit the sun in 1900 and use that as the standard year from which we calculate the subsequent divisions of seconds, minutes and days, we are creating an artificial picture of a year. Each year is unique, some shorter than 1900 and some longer. The point is that time as we use it is just a mental tool that we have adopted for organizational purposes.

We have incorporated the tool of time so thoroughly into our lives that we forget we are creating it in our minds and give it the status of an external reality to which we have to conform. A child does not come into this world with an understanding of clock time. It is an acquisition which he has to pick up in order to function in society. With adults always talking about this thing "time," and making such a big issue of

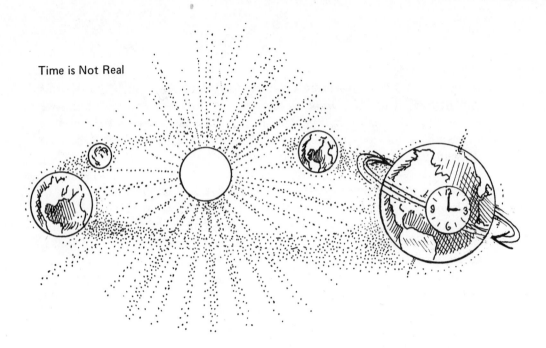

Time is Not Real

it, the child wants to be in on the game they are playing. Eventually the game is given the status of an objective reality against which to measure our experiencing and we forget that it is a tool to make life easier, not harder.

When we consider how many such tools are sitting inside our minds it is no wonder that most of us think of ourselves as separate entities in space-time. One of the major characteristics of the mind is its inclination to fix living processes in time, attach a label to them and pigeon-hole these processes into some concept *about* them. "Mind" has a very hard time experiencing life directly, outside its conceptual frameworks. This is the exact opposite of the new baby who is exposed to a deluge of strange sense stimuli and whose world is changing at an incredible pace. In fact it takes the baby a while to identify with the sensory organs as extensions of himself. We have all had this experience of seeing *through* our eyes rather than *with* them. We all had to learn to use our eyes, so why not unlearn in order to get back in touch with the source of all that we see, the seer within?

THE INNER SEER

Actually all skin absorbs and reacts to light, not just the eye, and the purpose of the following exercises is to remember how to see with our mind's eye, without our eyes.

Those light frequencies that we think we "see" as colors and brightness are actually virtual images created in our minds in response to patterns of electro-chemical reactions triggered by frequencies which resonate with our biological tissues. Color and brightness are purely psychical phenomena within our consciousness. Our experience of color therefore provides a wonderful map for the structure of our personalities and, as laid out in *Nuclear Evolution* by Christopher Hills, the key to our evolution.

To see without eyes means focusing the attention of the psychic seeing faculty in a piece of skin other than the retina of the eye, such as the skin of the palm of the hand. Blindfolds are necessary so that the level of light passing through the eyelids is reduced, otherwise the red of the blood in the eyelids is seen. Start by choosing two very different and contrasting colors, like yellow and red or indigo and orange. The colors can be squares of dyed felt or colored paper. The experiment should be done in a lighted room so the color absorbs and reflects light quite readily.

Discovering the Inner Seer

Hold your hands an inch or two over the surface of the colors and look for a subjective sensation in yourself. The difference between the two colors might be "seen" as a rapid tingling in one hand versus a coolness in the other, or an impression of darker dark over one color as contrasted with the other. You may even hear the colors, perhaps a high clear note for one and a ponderous, heavy tone for the other. Some colors might seem smooth and refined while others might seem coarse or sticky. If necessary touch the surface, as this is often easier.

When you can "see" a difference, move the two colors around so that you don't know which is which. Try to see the two colors and keep checking whether you're right or wrong. Include a friend in on the fun by having him move the colors for you. By frequently repeating this exercise you can quickly learn to see with your skin!

LEARNING A NEW LANGUAGE

Keep repeating this procedure with as many colors as possible. First "see" what they are like when you know what they are so that you have an awareness of what to look for when you don't know what they are. This evolves a color "language" that can be used with ever increasing discrimination. After several of these sessions, this language will actually start to manifest as seeing colors inside the head because the psychic visual faculty that interprets the electrical signals from the eyes will soon learn to interpret the electrical signals from any other piece of skin. One system becomes directly associated with another and you start to see with your skin!

Two of the most different colors are black and white. When all the colors can be seen with the skin the discrimination between black and white is fairly easy. Start by blindfolding your eyes and asking a friend to draw a large simple black shape or geometrical figure on the white paper. Once you have found with your finger some black on the paper, try to follow the line around. Ask a friend to tell you whenever you go wrong. By constant feedback the moment a mistake is made, there will soon be no need for the friend because you'll also have learned from the mistakes. Try smaller and smaller figures. Make them more complex. Try the letters of the alphabet. Cover the surface with cellophane so that there's no doubt that you are reading by feeling the ink colors on the page.

A QUESTION OF BLINDNESS

Isn't it amazing then that in spite of our skin vision, such a societal stigma is laid on people whose eyes do not work properly? A blind person is capable of actually seeing color, shape, and spatial position, yet is brainwashed into accepting a negative self-image that he is handicapped. In fact, in terms of seeing other realities like the emotional condition of a person or who he really is beneath all the fancy appearances, the 'blind' person often sees much clearer than the person with working eyeballs. In such situations being 'blind' is more of an asset than having eyes that work because these subtler vibrations are not being jumbled up with all the optical signals and it is easier to concentrate on the other's inner being.

DISTANT SEEING

Some people can eventually work their way up to perceiving distance and localization without eyes. Afer you have learned to distinguish the shape and color of a particular object, such as an orange, try keeping track of its motion as it is moved around your body at a distance of six inches. Ask a friend to start in front of your nose, slowly move it in some direction, and then stop. Sense where it is and get feedback from the friend. Continue practicing this, all the while increasing the arc travelled around your body until you are able to keep track of the orange anywhere around your body.

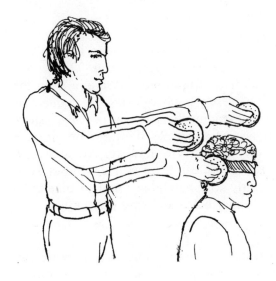

Have a friend blindfold you and move an orange to different locations around your head. Practice seeing it change position without your eyes. Start at a distance of six inches and gradually increase the distance.

Next try different objects until you are able to distinguish which object is where around your body. The minute doubt creeps in stop the exercise and go do something else. Patience will eventually pay off! The path to success lies in an attitude of openness to any sensations, and in not having expectations or thought patterns scrambling your receptivity. The next step, once you have learned to sense an object near to your body, is to practice sensing the object as it is moved to different locations within the room. Some people can actually see the color and shape of an object like an orange moving, even though it is behind them.

Sensing through uncovered skin is easiest so please don't get discouraged if these exercises aren't working when you've still got your overcoat on. Instead, take it off and try again. (Adam was probably much better at inner seeing before he ate the apple.)

A CLEAR MIND

The hardest and subtlest preliminary step that one who wishes to develop subtle sensing powers must master is stopping all the mental tapes to have a completely clear mind. Remember that because seeing is a mental process of interpreting electrical signals, it is crucial to make sure that what we're processing in our minds is a signal from the skin sensors rather than an illusion or imaginary conjuration having no correspondent in the physical world.

So learning how to still the mind of all its old movies is really the essence behind all these exercises if we want to succeed. The human mind is actually a receiver of countless vibrations coursing in from the cosmos and what we call "real" is merely a narrow range of frequencies which we are able to pick up more readily than others, just like a short wave radio's "reality" is short wave frequencies. Portions of this vast spectrum which we aren't normally sensitive to, however, are accessible with proper training, and this is why we can learn to see without eyes, see auras and even see clairvoyantly. Ultimately we learn to tune in to the subtle urgings of our deepest inner being and find out why we have come to live on the Earth.

The rest of this chapter is devoted to three examples of how we can tap into the profound transformative power within our imagination.* The first is an exercise in making sleep more refreshing, the second is an exercise in the use of the Tibetan symbol of cosmic balance, the Dorje, and the third is an exercise in clairvoyance. The sleep exercise is a simple

---

* For a deeper discussion of this see "Nuclear Evolution" by Christopher Hills.

technique that uses the imagination to deprogram and reprogram certain behavior patterns. It makes use of our power to create vivid virtual images by showing us how to imagine our bodies feeling very refreshed. The Dorje exercise is an example of how symbols can be used as creative gateways into powerful universal principles: a symbol can be used to create a virtual image which then acts as a bridge for contacting the invisible real image it symbolizes. Before we can be successful with the use of symbols we must first master the uncontrolled flow of virtual images which is presented to our awareness by our subconsciousness. Mastering this subconscious process is a situation like gardening where many of the virtual images are weeds and the real images are the flowers we want to blossom. If we do not clear the weeds away they will choke out the flowers and we will be left with a confusion of tangled subconscious images. Finally, the clairvoyance exercise offers a way of bringing our flow of subconscious virtual images to light so we can begin to see how they mold our conscious perceptions of reality and learn how to discard the images that are choking our true being.

THE POWER OF INNER SLEEP*

Set aside fifteen minutes before going to bed to practice this exercise. Sit in a comfortable position with a straight back, take a few deep breaths and feel the tensions built up during the day's work begin to melt down through your body and into the ground. Close the eyes and, starting with the tips of the toes, imagine going on a long journey throughout the body. At each stop along the way talk to the cells -- the blood cells, skin cells, bone cells, nerve cells, all of the cells -- and let them know how much their tireless work is appreciated by the boss, the consciousness which is expressing itself as a human. Relax with the cells at each stop and tell them that the whole body is going to be quiet for a few hours so that all the workers can relax and get refreshed. Tell each one that the big boss himself will be paying each a special visit to share the soothing silence with them. Imagine the cells feeling very relaxed and worry-free upon awakening. Slowly repeat this communication with all parts of the body from legs through the hips and torso to the arms and shoulders and finally up to the head.

As the head is relaxed imagine the consciousness being drawn up from all parts of the body to a deep, still point in the center of the head. Imagine this point as a bright star inside the head and see it actually radiating soothing rays that refresh whatever they touch. As these rays touch the insides of the head feel them caressing all the brain cells and the eye, ear, nose, and tongue cells. Feel the scalp cells slowly melt down into the soothing star quietly glowing in the middle of the head.

* This is an abbreviated version of the original exercise contained in Section Three of the "Into Meditation Now" course by Christopher Hills.

Next begin radiating these calming rays throughout the whole body and imagine the various cells becoming calm and peaceful as soon as they are touched by them. Soon the entire body will be filled with these soothing rays and it will be glowing with a calm radiance. Feel the whole being refreshing itself with each passing second it is bathed in this light. Imagine that as the body is sleeping it will be bathed in these special rays so that when it wakes up it will feel more refreshed than ever before. Imagine how it feels to be truly refreshed and ready to meet the new day with a sparkle of anticipation.

Finally imagine that as you practice this exercise each day the body will be refreshed and calmed increasingly, so that as you continue, your sleep will improve each night. Eventually you might find that you will not have to lie in bed as long as before yet you will feel even more refreshed!

DORJE: SYMBOLIC GATEWAY TO HEALTH

Dorje is the Tibetan equivalent for the Sanskrit word *vajra* which means diamond. Dorje refers to an essence that is so pure that nothing in the universe can affect it except itself, much like a diamond which can not be cut by anything but another diamond. Some insights into such an essence are triggered by noting the properties of neutrinos, which pass through matter as though it were transparent. Neutrinos are not affected by the electro-magnetic force and consequently carry no charge. Even though they are still particles of matter they can pass through matter because they are not reacting to the electric charges of other particles such as electrons and protons. The neutrino is neither positive nor negative, it is exactly the balance of the two polarities. Dorje expresses the essential unity underlying polarity and duality for it is in the relationship between opposites and in their union that the diamond-like qualities arise. There are certain schools of Buddhism called collectively "Vajrayana" (yana meaning vehicle or body, thus "diamond vehicle") that place this teaching in the center of their religious life. The teaching of the Diamond Vehicle is to discover the radiant diamond of the enlightened mind within one's own heart. The adept is called "vajra-dharma" or "wielder of vajra" and becomes a living synthesis of duality (lightness and darkness, form and formless, material and immaterial) perceiving the infinite in the finite.

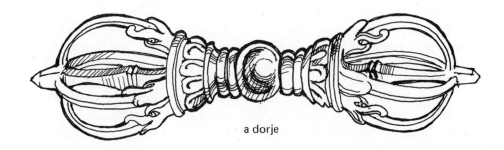

a dorje

As a three-dimensional symbol Dorje takes the shape of a sceptre, the emblem of supreme sovereign power, which we will refer to as "dorje" with small "d". In the center is a sphere which symbolizes the undifferentiated form of the universe. In Sanskrit this is called 'bindu' which means drop, seed, zero, smallest unit. Its potential force is represented by a spiral issuing from the center of the sphere. On two sides of this sphere are lotus blossoms representing the polar opposites (male-female, yin-yang, light-dark, good-evil, positive-negative). These petals face away from the central sphere, thus the sphere is also described as the jewel in the heart of the lotus. From these two lotus blossoms radiate spokes which converge again into a point at either end of the sceptre. This represents the differentiation of the original energy into many levels of energy just as the pure light of consciousness is differentiated into colors and the levels of consciousness. The differentiated energies then converge upon a point of higher unity forming the two ends of the symbol, the unified field of consciousness.

We are the artist who casts the design and the viewer who sees the form simultaneously, the creative and the receptive intertwined inseparably. The form that we perceive in our imagination is a virtual representation recreated from the piece of metal, which in turn is an artist's symbolic representation of the real image of Dorje. Unless they are consciously united, however, we continue to separate the one who sees from the one who is seen.

If we think of our imaginations as film studios where we are making movies *about* life, we can see that the movies we are making only represent our own points of view and are in no way the only perspective. While these movies are fascinating representations *about* the world we still have to leave the theater and live in the real world. We can gain new perspectives and insights but they are only worthwhile if we can make them come true in our direct experience of life.

Dorje then is not a "thing" we hold in our hands but rather a fundamental aspect of reality, a real image. The form that we touch, the dorje, is actually designed to be a guide to help us tune our imaginations to the Dorje principle, balance. By jumping out of our usual subject/object point of view into the unified field of consciousness where we experience Dorje directly without any separation, we discover that ultimate balance point, our true self.

USING THE GATE

The secret of the dorje is in its form and continued contemplation and wrapping of one's consciousness around that form will eventually reveal what is behind it. Dorjes are fairly difficult to obtain; certainly old ones that have been passed from lama to lama over the generations are very highly valued. As its value lies in acting both as a symbolic gateway into a state of perfect balance and as a receptacle for the energies impressed into it by preceding users, it is clear that a dorje is a very subtle tool. A dorje not only symbolizes a great power but also represents a chain by which the adamantine quality is transferred and added to by succeeding lamas. A dorje acts as a three-dimensional mandala and can be used as an object of concentration where one's attention becomes so focused upon the form that the false dichotomy of subject/object gradually melts into the unified field of consciousness.

The traditional method of using a dorje is in conjunction with a mantra such as "Om Mani Padme Hum" (Pure Consciousness is the Jewel in the Heart of the Lotus), which acts as a powerful aid in focusing consciousness into a vibratory pattern in resonance with the Dorje. As this is chanted the dorje is held in the right hand and the thumb of the right hand is rubbed along the length of each prong by revolving the dorje clockwise in unison with the rhythm of the mantra.

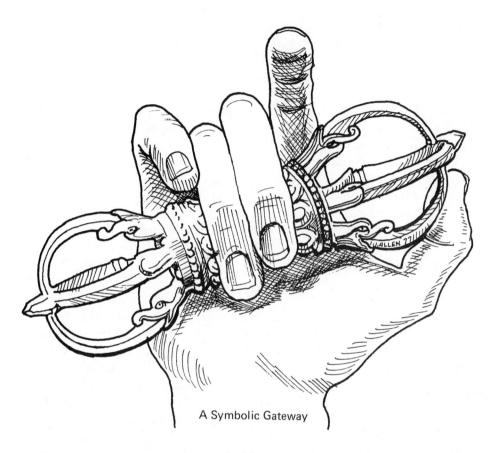

A Symbolic Gateway

Each prong is symbolic of a different aspect of consciousness, the central shaft representing Pure Consciousness and each of the prongs an aspect. If we are using a four-pronged model each prong can stand for one of the four states of matter: plasma, gas, liquid and solid, while if we are using an eight-pronged model each can represent the seven levels of being: red, orange, yellow, green, blue, indigo, violet and black. So as we rub each prong we tune in to each level and how it is balanced together with the others into the unified field of Pure Consciousness.

Once we have rubbed all the prongs on one end we turn the dorje upside down and repeat the process, tuning in to the nature of polarity. A positive would not be a positive without a negative and there would be no balance in the universe if there were not any opposites to balance. As we rub the opposite end of the dorje we can use our imagination to tune in to the marvelous interplay between opposites happening all around and through our consciousness. This marvelous interplay between polarities is the heart of sexuality. Everything in the universe is the result of a sexual union at some level of consciousness and we can use this exercise to tune in to different levels of sexuality at work in the cosmos, from interpersonal and international sex to interplanetary and intergalactic sex.

This Dorje exercise can be very valuable for cutting through knotty human problems such as interpersonal conflicts. More often than not gaining insight into why a conflict arises is the result of asking a key question and the reason that insight eludes us is often because we do not know what to ask. By going through the procedure outlined above we can take our consciousness on a ride into the cosmic balance point from which we can often spot the conflict-causing patterns of relating, which are not so obvious when we are caught in them.

AWAKENING OUR INNER VISION

A phenomenon of cosmic communication which transcends time and space that is closely related to telepathy is clairvoyance. While both faculties relate to sensing at a distance, clairvoyance means the perception of objects, events or people at a distance rather than the perception or reception of another's mind at a distance. Clairvoyance literally means "clear seeing." What is seen, though, is seen directly in the mind, whereas in normal vision the mind interprets the electrical signals of the eye. Clairvoyance, whereby the mind interprets some other signal, is in many ways similar to one half of telepathy -- that is, the function of the receiver. There are many reports of spontaneous clairvoyance, such as seeing the bus that one wants before it has appeared from around the corner, or picturing how something was or will be. Clairvoyance at will, however, has commonly been the role of such people as fortune tellers, crystal gazers, and others who use these methods to make small sums of money. However, if we recognize that it is the mind that sees, then such feats as tea leaf readings can be understood and used for what they are. There are two basic ways in which our faculty of inner vision can be used: (a) to see images arising about our own inner being, and (b) to see images about other beings. In the first lies a real key to health.

Any incongruency between the self-images of who we would like to be and who we are leads to inner conflict and disease, so it is important to be aware of our inner images because they reflect who we actually are. Throughout the day there is a constant flow of images passing briefly into our conscious mind, which often have nothing to do with our physical experience. Why do they arise? In our subconscious mind we have many hidden motives that lead us to make certain choices and all our conscious perceptions and images are heavily influenced by these motives. In order to get in touch with these motivations techniques have been developed which help a person project subconscious thoughts and desires into conscious images. The Rorschach ink-blot test is a well known example of a "projective test" where a person is shown an inkblot with an ambiguous pattern and is asked to tell the psychologist what images he sees in the inkblot. Another good example is mirror-

Rorschach Inkblot Test

Cloud Gazing

gazing. By concentrating on the surface of the mirror and purposely imagining a color on it, soon other distinctions such as contours, patterns and movement will follow.

There are several techniques called *scrying* used to induce clairvoyance and all of them basically consist of concentrating visually on something. The concentration serves to distract the conscious mind and this then allows the imagination free expression of subconscious experiences. Equally important in these techniques is relaxation, as it is only when relaxed that a person feels free to express openly.

Other variations of scrying include looking at the patterns in a pool of water, into sheep entrails, into a crystal, into tea leaves, and into a bed of hot coals. All these patterns could mean anything to anybody and therefore whatever image is seen is a projection of the seer's inner world into the forms. Such images clue us into our subconscious reality, our underlying experience of what reality is like. These are feelings which often do not agree with our conscious images and are therefore suppressed. Once these hidden drives are open to our awareness and we can acknowledge them as part of ourself we can integrate them into our conscious self-image. Such clear seeing and self-honesty gives us direction in our self-development, chops away the feeling of entanglement that comes when we suppress our subconscious desires, and leads to inner balance and health.

We can also pick up psychic impressions from another person, an object or event and actually see inner visions of these other realities. By tuning in to an object such as a bit of stone from an ancient tomb, a lost person's possessions, or a dorje passed from lama to lama, we can pick up psychometric images relating to situations the object has been in and tune through time into the past or future and tune across space into distant lands and cultures. This information may be recorded holographically in the crystalline structure of the object. Through daily practice of the following exercises we can learn to "feel" our way into the vibratory pattern of the object we are concentrating on and to receive images which resonate with the vibratory signature of that object.

The method of inducing clairvoyance through scrying is the same for all techniques. If, for example, a mirror is used, you would begin by imagining the surface of the mirror as being colored. Color is usually the easiest of all images to see. Black mirrors are used more than silvered ones. Once color is seen then all the other types of image will soon follow. When you are proficient in this, project an image of a simple object onto the surface of the mirror, then animals and then humans. With practice, images of people and events can be clearly seen. After this success you can tune in to the details of a person's life situation. We need to learn to firmly refuse any pictures that are not wanted, for this practice binds an image or loads the mirror with the power of imagination.

Step One -- Several different methods can be used by a group interested in working together because different people will be effective through different methods. Some suggested materials to have are: a crystal ball or any other shaped crystal placed in water, a mirror, a piece of glass painted black on its back side, a cup or glass of liquid, running water, basin of water, stones from beach or river bed, one hand with soot and oil, a sky full of clouds, and if possible a spring (of water). When each person has decided which method he or she will use a short description of each technique is read out.

Crystallomancy -- is now the most common method of scrying and the name crystal gazing speaks for itself. Originally the crystals were put in water as water was believed to have some magical property. With the popularity of the crystal ball this practice has ceased. *Catoptromancy* -- is sometimes called enoptromancy, and means using a mirror or flat piece of steel on to which mental or imaginative scenes are projected. This method was widely used at one time in primitive tribes of Central America. *Cylicomancy* -- is the use of a cup of liquid or glass of wine as the focus of attention. *Gastomancy* -- now covers two different kinds of scrying. The original method was to use the marks on a human stomach for scrying. The term later came to mean the use of belly-shaped bottles for scrying. *Hydromancy* -- uses rivers, lakes, streams, etc. for scrying. In this session a basin or a bath full of water could be used. *Lecanomancy* -- is very similar to cylicomancy, but instead of a cup a basin is used. *Lithomancy* -- is using large pebbles or stones of the kind found on beaches that have many interweaving patterns. Alternatively, semi-precious stones could be used. *Onychomancy* -- was originally used by taking a young virgin boy and covering his hands with soot and oil, and then turning his hands towards the sun, the images that appeared representing the answers required. The next method has no name, as far as I know, so perhaps one can be invented, but it consists of looking at the changing forms made by clouds in the sky, the marks in wallpaper or the patterns in the bark of trees. *Pegomancy* -- is a specialized form of hydromancy which uses the scrying of water from springs.

Step Two -- Before starting, a short exercise is done with each object. In order to tune in, and to fully concentrate, the first five minutes should be spent carefully examining all the small details of shape, colour, texture and even taste. The five minutes should be spent without one moment's lapse of concentration. After five minutes Step Three is read out, and then scrying can begin by allowing the object to form the basis for the images.

This should take about 10 minutes. A discussion follows where each person describes his or her images and their meaning as briefly as possible. Notes should be made to compare what is seen with what actually happens. Refreshments are served, and tea is included; then naturally the tea leaves should be left at the bottom of the cup (if coffee, then leave the dregs). These can be used after the initial experiment, each participant reading his own tea leaves and noting the images that arise according to Step Three.

Step Three -- The instructions are very simple. Each participant simply becomes aware and remembers all the images that are seen in the mind. If he decides to see what is happening to someone at a distance then he can mentally concentrate on that person and see what is seen. Similarly for the future, if he decides to see what will happen next Thursday, then hold in attention that decision, and wait for the images to appear. Some of the images will be realistic and some fantastic. Make a note of them all; the fantastic ones will not be so fantastic when they actually happen.

Step Four -- If someone loses or forgets images because of a difficulty with concentrating, he or she can find help in the concentration exercises in Section One.

THE MAGIC MIRROR OF LIFE

Our imagination can recreate anything we want regardless of good and bad, so whatever we experience as reality is in fact a mirror of how we are using the imagination. We all have the same imaginative power but we all put in different programs and each of us creates a different reality for ourselves. How we program our imagination is crucial; how do we know for sure the program is healthy?

The answer is in learning how to use the magic mirror of life itself. Since whatever we experience is a reflection of our self-programming, we can use our experiences of life as clues about our self-images. If we experience a lot of pain and misery and nothing seems to be going right then this is showing us that the film studio which is our imagination is stuck in a rut of producing tragedies. If life seems like a big sham then that is what we are producing. If life is fulfilling and we feel content, then that too is the kind of movie we are creating with our imagination.

But there is something beyond all movies, some intelligent purpose in the cosmos which has created this whole show, including the imagination. If we can learn to use our imagination to contact this purpose then we begin to leave all our mental tapes and programs behind like a butterfly shedding its cocoon, and break through into a whole new dimension of being where there is no experience of an experiencer experiencing a "world out there" because there is no separation between knower and known, subject and object. In this state the constant games of creating mental models and status quo and striving for social prominence as a special individual are left behind for an entirely new relationship with life. It is the difference between the young child who is constantly wonder-full and does not "know" anything and the adult who is beset with cares and worries and clogged up with all kinds of socially acknowledged information. If we look in our magic mirror which one would we want to see staring back at us?

This is the theory, that we have the power to recreate the world in our image and can therefore free ourselves from four-dimensional thinking trapped in space-time. It is easier said than done. The real work lies in making this theory a living practice.

Chapter Ten will take us one step closer by introducing *supersensing,* a generally unknown faculty of perception lying dormant in all of us. In using it we automatically step beyond the limits of space and time into a dimension of being which can be anywhere, anytime; the domain of *Supersensonics.* We have seen that the crucial factor is *what* we create. When we create a self-image of a being who is limited to a body our experience of the cosmos is limited accordingly, and when we create a self-image of unlimited consciousness radiating in all directions our experience will begin to reflect that too.

SUPERSENSING

We each live in a world of psychic holograms which could very easily be completely distorted representations of reality. How do we know what is real? We need a reliable method for validating our perceptions about life. Empirical evidence is not valuable here because our experiences are also images created in our minds and subject to distortion. Logic is useless because the basic assumptions cannot be verified. A completely new approach is needed, like a hotline to the Creative Intelligence.

Long ago in ancient China, Egypt and India, researchers used an amazing supersense which up to now has been largely neglected by established science. Through it they were able to map the circuits of consciousness at work in creating psychic holograms and validate their perceptions. The ancient flowering of civilizations in China, Egypt and India all proceeded through the use of this divining supersense and great masterpieces such as the *I Ching*, the Great Pyramid and the *Yoga Sutras* were developed with knowledge gained through this faculty.*

Though it is dormant in most of us, we all have this supersense and with proper training we can easily awaken it. Our mental images of material forms only acknowledge those vibratory frequencies which our physical senses are structured to interact with, yet on either side of that narrow slit of awareness stretch much vaster ranges of vibratory frequencies which our supersense can detect. Even though thoughts, emotions, atoms, black holes and a host of other things are all vibrating waveforms nonexistent as far as our senses are concerned, with our supersense we can tune in directly to them and experience their inner reality. New dimensions of reality beyond time, space and matter will open to us through supersensing once we learn how to use it.

RADIESTHESIA: THE SUPERSENSE

Supersensing is often referred to as radiesthesia, dowsing or divining. Radiesthesia means sensing radiations. The term was coined in the 1900's by Abbé Bouly, one of the group of French pioneers in biofeedback including Louis Turenne and Henri Mager. Radiesthesia goes back into

---

* See "Supersensonics."

antiquity as a method for finding water and precious minerals. Basically, radiesthesia is a form of biofeedback and is commonly practiced in many countries in the form of dowsing for water and minerals. Water dowsers first fix their thoughts on WATER and then systematically scan the territory they are prospecting until their nervous systems detect radiations which resonate with the radiational "signature"* of water held in their consciousness. Because the human body contains so much water it naturally resonates with the water a dowser seeks and hence is one of the easiest signatures to detect. When it comes to minerals such as oil, however, dowsers often hold a vial of oil because it is much easier to tune their nervous systems to the actual physical sample than merely to hold the image of it in their minds. Such samples are usually referred to as "witnesses." Once the nervous system picks up the resonant frequency there is an obvious neuromuscular reaction. The minute muscular movements are then amplified, by a dowsing rod for instance, into very visible dips in the direction of the water.

With a good dowser the chances are 100 percent that where he or she says to dig, there will be a pleasant surprise at the bottom of the hole. Top geologists consider themselves lucky to be 50 percent accurate in

* Signature: the unique vibrational wave field by which a substance can be identified.

predicting water, and oil drillers 25 percent in predicting oil. The diviner's rod can give a massive reaction or become very subtle in its movements. A spring tension added to the dowsing rod amplifies the very subtle neuromuscular responses.

CONCENTRATION AND RECEPTIVITY

The keys to successful dowsing are concentration and receptivity. By *concentration* we mean the ability to focus the mind exclusively on the vibratory pattern being dowsed for and by *receptivity* we mean the ability to make the mind so still that it is open to the subtlest resonant vibrations. Trying to do radiesthesia without proper concentration and receptivity is like trying to listen to a favorite radio program without a radio set. Concentration is needed to pick out the right vibrations from the deluge bombarding our neural antennae. Receptivity is needed in order to put aside all the mental tapes that usually play so that they are not causing interference noise which drowns out the message. Actually, radiesthesia is somewhat similar to radio technology because in both fields the operator is tuning the receiving equipment to pick up the resonant signals and disregard the irrelevant "noise." There are many techniques for cultivating these skills but the essential ingredient is total commitment. A half-hearted approach will only bring half results and this will end up disheartening the student.

The sensitivity of the operator reaches a saturation point easily, so operators are encouraged to work only during times of optimum sensitivity. Sunlight makes dowsing much easier, so whenever possible practice in direct sunlight. As found in telepathy experiments sensitivity or receptivity also decreases when the person is stressed or anxious, during thunderstorms, etc. Any large quantities of electromagnetic equipment or machinery (steel, etc.) nearby will also be disturbing. Avoid experimenting with too many people around because if they are skeptical they will be sending out thought waves which will tend to negate sensitive experiments. Even one person with a negative wave field can quench the subtle radiations of material objects if you are performing a group experiment.

OUR FANTASTIC ANTENNA

The main difference between supersensing and other forms of perception such as telepathy or clairvoyance is that a radiesthesist uses the neuromuscular reactions to signals picked up by the nervous system to tune in to Nature while a telepath or clairvoyant uses mental imagery arising in response to these signals. The signal of perception is an unconscious involuntary neuromuscular reflex triggered by a resonance effect in the consciousness of the operator. The resonance effect happens when the signals received by the nervous system match the vibrational signature held in the mind. The neuromuscular response to this resonance is made accessible to visual perception via an amplifier such as a divining rod or a pendulum. It should be noted that the radiesthesist can gradually develop a sensitivity to the unamplified neuromuscular responses in the fingertips so that rods and pendulums are no longer necessary. And further, if the student really wraps his/her consciousness around training this sensitivity, eventually the signal of resonance will be experienced directly in our consciousness with no need of a neuromuscular tingling in the fingers.

So the crux of radiesthesia is not the amplifiers, it is the operator's own consciousness. The tools are only for amplifying nervous signals; training the natural biofeedback in the neurological circuits inside our consciousness is the real job.

THE YES AND NO OF IT

The immense network of intertwined nerve cells which act as a communication system for our body is characterized by its on/off binary operations. It is elegantly simple. Expanding the picture from a single nerve's action or nonaction to the vast neural networks which are at work, let us consider the situation of a symphony orchestra where each instrument acts at different intervals and the communion of them all results in a symphony. The interval between actions for each instrument is crucial; vary it slightly and you no longer have a symphony but a cacophony. In the same way the nervous system uses precise arrangements to orchestrate the body processes into a symphony; illness is merely an example of cacophony. In the case of radiesthesia the vibrational signature of the witness determines the signature or key in which the nervous instruments of the orchestra will play and the whole nervous system is primed to resonate with any vibration in the environment having that signature.

Because the nervous system is a binary information distributor it will either respond to the signals being fed into it or it won't. This binary on-off set-up is carried through succeeding orders of organization to the neuromuscular level. This can be seen most easily in the up/down movement of a dowsing rod or the clockwise/counterclockwise rotations of a pendulum. In other words, the neuromuscular responses are amplified into binary movements which tell the dowser either "Yes, there is water in that direction" or "No, there is no water in that direction."

The radiesthesist can discover through movements of these amplifiers the same information that the telepath perceives directly. The major advantage of radiesthesic methods is that the binary language of radiesthesia enables more abstract information to be acquired in a scientific way about the many different aspects of energy itself, as well as the nature of energetic rhythms and patterns. The telepath acquires perceptions in terms of images and symbols which contain a huge amount of information which then needs to be sifted for its grains of truth. These images are also unique to the person experiencing them, that is, they are not replicable because each person has his own symbolic language. On the other hand the information acquired via supersensing is in the form of yes/no responses to particular questions which can easily be duplicated by any radiesthesist because we can all get the same answers: up/down or clockwise/counterclockwise. This immediately makes supersensing more valuable to the scientific method because we can test existing assumptions by this form of questioning.

By "scientific" we mean that a logical method can be used to discover the nature of life processes as compared with the intuitive leaps of the clairvoyant. For instance, let's imagine ourselves in a garden infested with slugs which we want to get rid of. With supersensing we can develop a logical series of yes/no questions which will eventually lead us to an effective solution. With clairvoyance we might immediately see the imbalance in the bioenergetics of the garden system which was encouraging the slugs to feast.

In practice we have found that supersensing is valuable as a tool and and like any other tool must be used in conjunction with other methods of investigation, be they psychic or scientific, as it provides us with a way of validating our intuitions and theories through firsthand experiencing of vibrational interrelationships. Supersensing, rather than contradicting traditional scientific methods of investigation, complements and enhances them because it can raise the researcher out of the dry uncertainty of theoretical secondhand "knowledge about" something into firsthand direct experiencing from within. Instead of separating ourselves from life processes by observing from outside and making theories on what might be happening inside, we can actually establish a direct link with the life we are investigating. By tuning our nervous systems to resonate with that particular vibrational pattern we gain direct perception from within, as if we were actually inside that bit of life.

Naturally, this leads into deep considerations about the essence of life itself. Because we are only interested here in introductory aspects, we refer those who would like to delve deeper into the nitty-gritty physics and philosophy to the works of Christopher Hills, especially

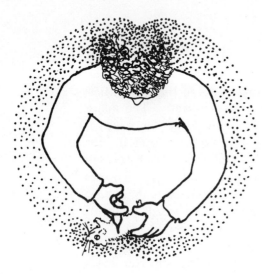 

The common scientific viewpoint of being separate from the object of study does not realize that both researcher and test mouse are bound together in consciousness.

Supersensonics. The promise of radiesthesia is that we can explore the universe directly and find out firsthand what we need to know, as long as we phrase our questions in yes/no terms. The answers are all in the senses of the beholder when we put them to work.

AURA BALANCING

A considerable amount of radiesthesic attention has been directed toward color and wave frequencies, as color is related to every vibratory form whether it be a physical solid, a magnetic field, or an aura. Using color as a language of vibrations, the early radiesthesists were able to organize their findings into a simple model. They found that the vibratory patterns of different substances could be classified according to their resonances with the seven colors of the visible light spectrum. Proceeding from this basic finding they eventually discovered that color could be used as a basis for biological diagnosis.

A good example of its application in diagnosis happened recently with a physicist, Robert Massy. He was visiting with an M.D. who happened to mention that he couldn't bend over and touch his feet because of severe pain, presumably having to do with the sciatic nerve. This doctor (referred to here as "L") asked Massy if he could use radiesthesia to help outline the source of the problem. Agreeing to try, Massy scanned L's body with a pendulum until he pinpointed a spot in the small of L's back as a troubled area. Gently pressing this spot, Massy then asked L, "Is this the spot?" With a yelp of pain and amazement L remarked that the point had indeed been located. Next Massy used a Positive Green pendulum to draw out the twisted energy from that point and then asked L to try and touch his feet. L bent over and easily touched his feet! Knowing, however, that the physical symptoms were results of some deeper psychological problems, Massy then scanned the various chakras to find which one of them was most severely blocked. Green, the heart center, responded as the most blocked so he asked L if he was having severe emotional upsets at home. It turned out that L was still quite upset over his wife's death and had been in extreme heartache for years over losing her. Massy then proceeded as far as L wished to go with his situation at that time.*

CONSTRUCT A SIMPLE PENDULUM

The art of supersensing lies in being able to detect specific vibratory "signatures" and to block out the remaining signals, which for all practical purposes constitute noise. To this end radiesthesists employ two supersensonic methods: amplification and selectivity. By using specific shapes, materials and "witnesses" we can strengthen both the signal being received and the receptive capabilities of the receiver. Supersensing is similar to the situation of a stereo "receiver" which incorporates both an amplifier for boosting the signal strength and a tuner for picking up the radio waves. The pendulum acts as an amplifier and the shapes, materials and witnesses of which it is composed act as tuners to zero in on specific frequencies.

Aura Pendulum

A very simple pendulum can be made by using a plain wooden dowel approximately 3 1/2 inches long with a diameter of 5/8 inch. First make a point on one end. Next drill a tiny 1/16 inch hole in the exact center of the flat end. It must be in the center or the pendulum will not balance properly. Next insert a 6 inch length of heavy black thread into the hole with some epoxy cement. Let the cement set overnight.

There are many different pendulum designs for specific purposes.* For instance, the aura pendulum is quite similar to the simple dowel design we introduced above with the addition of a black stain and a spectrum strip down one side. This strip acts as a witness for the spectral differentiations inherent in all vibratory patterns, especially those related to the different levels of consciousness. By sliding a small cursor up or down the strip to a particular color we make the pendulum selective for that band of vibration so that if we move it to green, for instance, we automatically screen out other energy bands from interfering with our results. This particular design is for use in deter-

* A more detailed version of this story is included in the forthcoming book "White Magic."

* See "Instruments of Knowing", a catalogue of divining tools available from the publishers.

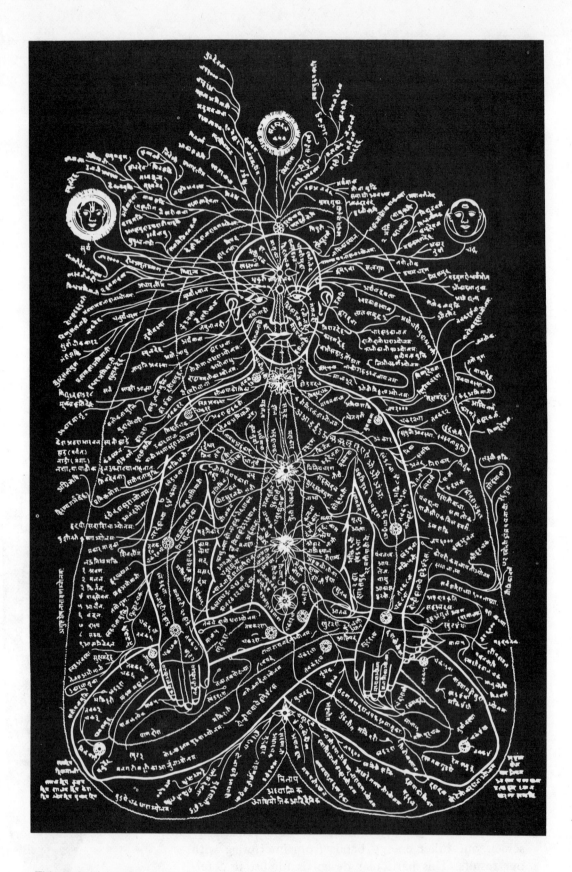

This ancient drawing by the Sanskrit seers shows the psychic electricity channels of the diviner's nervous system. The seven chakras act like filters for the sun and moon energies. The sun stands as a symbol for cosmic rays ionizing primary unpolarized light. This affects the right side linked with the sympathetic nerves. The lunar rays stand for reflected and polarized light mirrored from the objects of creation linked to the left hand side and the parasympathetic nerves. The diviner's neuromuscular reaction is a transduction from psychic electricity ⟶ nervous signal ⟶ muscular chemical signal.

mining the relative functioning of the various levels of human consciousness within us via our auras. The black coloring of the pendulum enhances its absorption qualities, as black is a color which readily absorbs radiative energies of any kind. So between these two characteristics we have a pendulum much more sophisticated than a mere wooden bob. At this point we need to reiterate that the pendulum is not "doing" it, it is only a wave guide for the consciousness of the operator. The thing that is "doing" it is the consciousness of the operator in combination with the tool. It is the same with a computer. The computer is limited to the intelligence of the programmer or its user. Like any other tool, it can often do more than it was designed for.

Once a pendulum has been made or obtained, the next step is to tune it. Because a pendulum is an extension of the operator's nervous system, it is essential that the length of the pendulum's string be proportional to the vibratory frequency of the neuromuscular impulses. It is the same situation as a bell tower where the ringer times his pulls on the rope to coincide with the apex of the bell's swing in order to take advantage of the inertia. If he pulls the rope before the bell is swung all the way up then the whole rhythm of the bell is upset and the force of its swing is drastically impaired. With a pendulum the variable is the length of the swing rather than the timing of the pull, which is automatically provided by the nervous system. The nervous system sends out a regular signal for triggering the arm's muscular movement so it is very important that the pendulum's arc be exactly timed to match. By finding the right length, the operator is tuning the pendulum to his or her nervous system. This is done simply by holding the pendulum in the right hand over the left hand, reverse for left-handed people.

Start by holding the thread between the thumb and first finger at the point closest to the pendulum. Swing the pendulum to and fro in a two o'clock to eight o'clock direction. As it is swinging, slowly let the string out. A point will be reached on the string where the pendulum starts to rotate in a circle. Continue to let the string out slowly and the pendulum will oscillate to and fro again until a second "tuning" point is reached where the pendulum starts to rotate again, but with much greater rotation than before. STOP! Here is the optimum point of tuning. If in doubt repeat the procedure several times until you find a tuning that feels right. You may wish to tie a knot in the string at this "tuning point" so you can find it quickly.

LEARNING ITS LANGUAGE

Once the pendulum is tuned, its own new language must be learned. Each person has his own language or code in that his nervous system reacts differently than anyone else's to the same signals. In other words, in response to the same vibrations one person's pendulum might rotate clockwise while another's moves counterclockwise. However, the language is basically simple as there are only four types of movement: 1) circles or gyrations, 2) ellipses, 3) straight lines or oscillations, and 4) spirals.

Information is gained, not from the actual movements themselves, but from the language or code in our consciousness. The gross movement comes from the movement voluntarily initiated by the person, but the direction and duration of movement comes from the involuntary neuro-muscular reactions to the vibrations received. The most simple code is a yes/no, positive/negative code. An example of this is the clockwise gyration for "yes" over the north end of a magnet and counterclockwise gyration for "no" over the south end which is common to so many dowsers. Whatever direction of rotation we get over the north end of a magnet will be our "yes" and whatever we get over the south end our "no."

Everyone discovers his own code; there is no right or wrong code. The code can be discovered by using an ordinary bar magnet. Around a straight magnet a diviner can detect three forms of the field: 1) a positive form around the magnet's north-seeking pole, 2) a negative form around the magnet's south-seeking pole and 3) a radio-magnetic form (+ --) immediately over the center of the balance point between polarities. The most common code is a clockwise rotation for positive, a counterclockwise rotation for negative, and a composite rotation over radiomagnetic consisting of a positive rotation followed by three north-south oscillations followed by a negative rotation followed by three east-west oscillations and then back to a positive rotation, etc.

An excellent introduction to practical radiesthesia is Robert Massy's *Alive to the Universe*, particularly in light of the fact that many beginners try to attempt much too sophisticated experiments right when they are beginning and, upon failing, get discouraged. The secret is to start out gradually with sensitivity exercises and gradually build up to the more skilled and difficult applications. Most people will spend considerable time learning to play a musical instrument but to learn to play their own instrument of the nervous system they usually want to be a skilled performer without any effort. Nevertheless it is considerably easier than learning the guitar.

THE DOWSER'S ROD

Most water dowsers use a tool called the "dowser's rod" for finding water. This traditional instrument can be made by taking two pieces of springy wood (hazel or willow works fine) and binding the first inch of each one together with wet nylon cord. After the cord has dried, coat

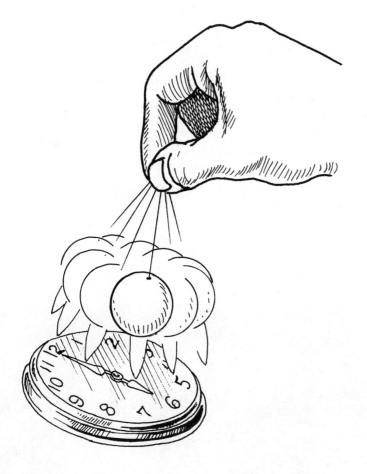

TUNING YOUR PENDULUM

Start by swinging the pendulum back and forth and letting the string out slowly. Soon it will begin to rotate. Keep on letting the string out. The pendulum will begin to swing back and forth again. Continue letting the string out slowly until the pendulum begins to rotate a second time. STOP. Your pendulum will be tuned.

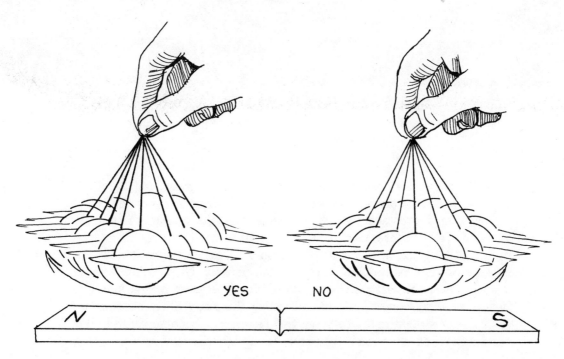

When tuned, your pendulum should rotate in opposite directions over the north and south poles of a magnet.

the binding with a generous coating of epoxy cement. These rods should be about eleven inches long and three-eighths inch in diameter. Also make sure they are both as similar as possible with regard to diameter and spring. Other material can be used such as plastic, whalebone, or wire.

When they are bound, separate the two ends of the rods by placing the thumb on the inside of each rod and the first finger on the outside, letting the rod rest against the next three fingers. The little fingers around the rod rest on the palm. Next keep the palms of the hands upwards but as flat as possible with the elbows held close to the body. The arms should be firm, yet relaxed and springy.

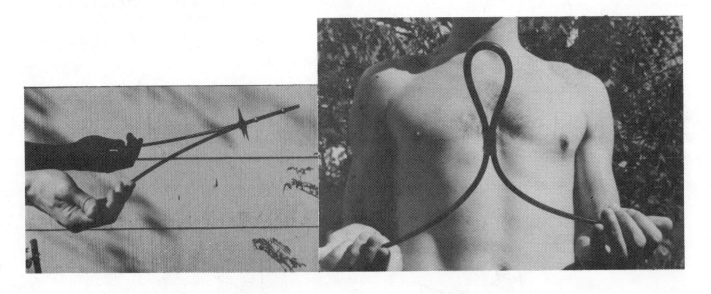

Turenne Magnet-fitted Rod

Egyptian Ankh

Thus the two rods are held apart at a constant tension and they will now rise or fall by pivoting on the fingers, depending on when the rod is in line with a substance which resonates with the operator's thought waves. As it takes practice to become proficient with a rod it is very important that the novice not go out on the first experiment with the expectation that the rod will rise and fall accurately. Nor should there be any negative expectations. Expectation feeling is good but it must be that of expecting anything and everything. Like any other instrument, the dowsing system takes patience and practice to become reliable. Big expectations at the start inevitably establish thought waves which interfere with the vibratory fields of the substances being dowsed for, so drop those big ideas and start humbly. Expect, but expect nothing in particular.

The rods pictured above were made according to a design perfected centuries ago by the ancient Egyptians. This design has become symbolized by the ceremonial Ankh or Key of Life which is generally associated with the dynasties of the Nile. About 3000 years before it became a ceremonial symbol of the Egyptian civilization, the Key of Life was a divining rod made from one piece of wood bent double and bound one inch below the bend, as can be seen in the illustration above. The cross was a way of representing the field discharged from the bound section of the rod when the operator found the resonance. The accuracy of this design is attested to in the marvelous engineering of the pyramids, designed with supersensitive instruments of knowing.

Another particularly effective rod is that designed by Louis Turenne. It incorporates the design mentioned above with the use of magnetic needles at the binding site. These needles allow the operator to make the rod selective to horizontal waves, vertical waves and east-west waves. Horizontal waves include magnetic waves, vertical waves include electrical waves, and east-west waves include those generated by the stream of free electrons which flows around the earth in an east-west flow. This rod is favored by adepts because of its selectivity.

MAPPING ENERGY FIELDS

Magnet-fitted Pendulum

The previous experiments with a magnet showed that there are three types of fields in a horizontal plane over a bar magnet (positive, negative and radiomagnetic). These can be detected to a height of eight inches above the magnet. Above this height and to the sides of the elements being tested a vertical influence can be detected, either by using a pendulum that does not react to horizontal fields (for example, a mercury filled pendulum such as the Merkhet*) or by fitting magnets to the pendulum in order to cancel out horizontal fields.

If you are using an ordinary pendulum, hold it more than eight inches above the center of a compass magnet and the north-south, east-west oscillations of this vertical influence will be detected. Turenne has called the alternating (+, −) force "radiomagnetic."

* See "Instruments of Knowing."

BUILDING A VOCABULARY

When you are proficient at detecting both horizontal and vertical type influences over magnets, a wider experience can be gained by testing yourself on as many different materials as possible. A few are given below as a check. (A zero indicates no influence.)

Element	Horizontal Influence	Vertical Influence
Carbon	+	+ –
Aluminum	+ –	+ –
Iron	+	+ –
Copper	+	+
Zinc	–	0
Silver	–	+ –
Tin	+ –	0
Gold	+	+ –
Lead	+	–
Thorium	0	+ –

ELECTRICITY AND MAGNETISM

Electricity and magnetism are directly related. This can be easily seen in the generation of a magnetic field by a bar of iron with wire coiled around it when an electrical current is passed through the wire. In fact both forces manifest fields at right angles to each other, magnetism on the horizontal plane and electricity on the vertical plane. Gravity waves may account for the east-west flow of electrons.

To examine the complex field produced by these energies acting together, a long electromagnet can be made in the form of a solenoid through which a current from a battery can pass. A thirty-two foot wire is wound so that the two ends are oppositely coiled (see diagram below). The wire is laid on the ground so that the end that attracts the south-seeking pole of a compass lies north and vice versa. The fields more than a very short distance away from the wire are very small so the sensitivity of the operator can be aided by using a small magnet as a witness for the fields being studied. By holding the magnet horizontal, for instance, it acts as a witness for horizontal fields.

In addition to the usual north-south fields around a magnet which can be seen by putting iron filings around it, the radiesthesist is now able to detect east-west fields flowing at right angles to the middle of the magnet. (See diagram following.)

By aligning a solenoid made of coiled wire with the earth's N-S magnetic axis we can easily generate electro-magnetic and east-west fields.

Early radiesthesists in the twenties discovered that in addition to the familiar north-south flowing fields associated with magnetism and electricity there were these unknown east-west flowing fields of ions streaming around all things, including plants, animals, minerals and the earth. This has led Christopher Hills to postulate that all matter is polarized in three planes around four poles.*

All matter is magnetic, paramagnetic or diamagnetic so if the filings which are magnetic are made into a suspended needle shape, they will tend to lie with their length parallel with the magnetic lines of force. If diamagnetic, they will lie east-west. This means that all matter has an east-west force field of energy, including our own bodies. [Note: a *paramagnetic* substance has a magnetic permeability greater than that of a vacuum and is positively attracted to a magnetic field. *Diamagnetic* means that the substance has a tendency to lie at right angles to the magnetic field as though it were being equally repelled and attracted by both poles. This is because it has a negative magnetic susceptibility or a permeability less than that of a vacuum (e.g. bismuth, antimony, or copper.)] The ancients discovered these relationships through dowsing and used them to amplify the supersense. E.g. antimony powder was found sprinkled on the floor of the King's Chamber in the Great Pyramid.

SYMBOLS ARE MAPS

If we concentrate intensely on a symbol it can act as a focusing lens to guide our consciousness into awareness of the dynamic life processes within the cosmos. With our supersense it is possible to map out definite wave fields emanating from symbols which resonate with certain cosmic forces and this allows us to use symbols as radiesthesic witnesses. The dorje exercise in Chapter Nine is a good example of this principle at work. This avenue of research can be elaborated endlessly to include symbols such as letters of the alphabet, numbers, signs of the zodiac, runes, and even geometrical shapes occurring in Nature.

* See "Supersensonics."

All our basic geometric shapes such as lines, circles, triangles, rectangles, squares, pentagons, hexagons and octagons are borrowed from Nature and act as links with the forces that create the natural forms. We can see the straight line in the tall stalk of a wheat plant, the circle in the rings formed in a pond when a small pebble is tossed into the middle, the triangle in the shape of certain mountains or ant hills, the square in a cross section of a mint stalk, the rectangle in the shape of a tiny microorganelle inside a plant cell, the pentagon in the five-sided arrangement of many flowers, the hexagon in the six-sided crystals of snow, and the octagon in the eight-sided construction of a spider's web.

The ancient Chinese sages were great students of Nature's forms and their symbols reflect wonderful insight into the use of forms as supersensonic links with natural forces. A particularly fascinating example of this is the fact that trigrams can be used as witnesses for natural forces. In *Supersensonics* Hills describes how the trigram CHEN ☳ (The Arousing, Thunder) can be used as a witness for oncoming storms and how in conjunction with a map the trigrams can be used for weather forecasting. Another example of the power of symbols is Solomon's Seal. Solomon's Seal, or the "Star of David," consists of two interlaced triangles whose apexes can all be circumscribed by one circle and a pendulum will show that the interior of this symbol is negative while the six surrounding triangles are positive. This symbol produces only vertical waves, it has no horizontal influence, so our curiosity prompts us to wonder if this seal can serve as a radiesthesic shielding from unwanted emotional fields.*

By placing a witness of ourselves inside the seal (inside the hexagon) before exposing ourselves to a highly emotional situation, such as the atmosphere of a stock exchange, we can test this hypothesis. After about half an hour of excitement we call an assistant and have him remove the witness (a photograph makes an excellent witness) and then note if there is any change in the effect on us of other people's emotional vibrations.

This also give us a new insight into the ancient tales of demons sealed inside copper flasks by Solomon's Seal. It is very likely that the "demons" referred to in this ancient allegory are the emotional turbulences we all have to become masters of if we are ever to achieve self-mastery and fulfillment. By placing these emotional demons inside a vertical wavefield it becomes easier to see what they truly are rather than being confused by them.

* The horizontal type waves which emanate from secondary or reflected light emissions are inversely proportional to the vertical waves emanating from primary sources of light. The fact that radiesthesists can relate horizontal waves to emotional states and vertical waves to mental states implies that magnetism (horizontal) and electricity (vertical) are linked with our psyche.

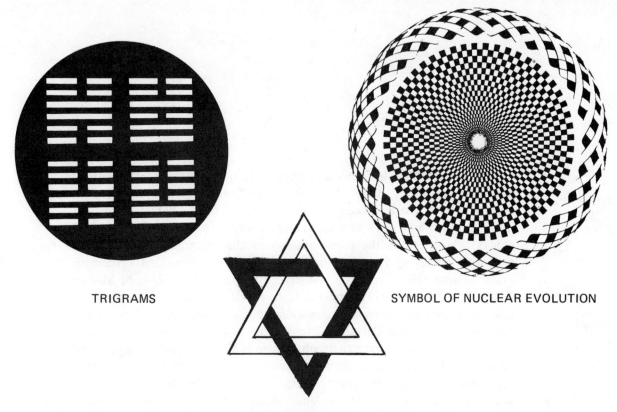

TRIGRAMS   SYMBOL OF NUCLEAR EVOLUTION

SOLOMON'S SEAL

Changing the orientation of the Seal with respect to north-south will have an effect on the field patterns, so experiment to find the most effective alignment.

COLORS OF LIFE

Since all material forms are actually crystallized light, it follows that any "laws" governing the diffraction of light into the seven-colored spectrum will have a direct relationship to other patterns of crystallization such as the elements in our bodies. Early in this century French pioneers in radiesthesia discovered that all crystallizations could be identified according to their "fundamental angles" of light absorption.

Taking a complete circle as the symbol for pure, undifferentiated light, they found that different angles would resonate with each color of the spectrum. This resonance was easily detected by the diviner's reaction via a pendulum. Eventually Turenne developed the Fundamental Ray Disc for analyzing chemical structure through application of these findings. By finding which angles a test substance resonated with he was able to know which colors it was related to. In turn, he also knew which angles resonated with specific elements of the periodic table and by finding the resonant angles of an unknown substance, he would be able to correlate these angles with the known angles of the basic elements. Interestingly, in the hands of a skilled radiesthesist, such as Turenne, this tool enabled the operator to be as accurate as or even better than the best analytical equipment devised by physics. Turenne himself was able to predict and detect the existence of several elements unknown to the conventional science of his day.

The explanation for this accuracy lies in the nature of scientific measurement itself. All the sensitively calibrated instruments used to detect minute changes (and so let scientists know what is happening inside life systems such as atoms) are based on mental yardsticks and clocks originating in the consciousness of the scientists. For instance, the meanings of "cycles per second" or "Hertz" are dependent on certain units of time called "seconds" which exist only in the minds of human beings for explanatory purposes. Nature has no need of seconds. In this way, complex units of measurement are constructed in the imaginations of all those undertaking the training to become scientists because in order to communicate their observations in terms of language, they must agree beforehand what meanings to assign to the descriptive terms they will be using. In other words, they all subscribe to generally accepted models of the universe, and the instruments they use to measure or sense the universe have the same conventional limitations of these underlying concepts about the nature of reality built into them. Any tool is only a refinement of the abilities of the user and this includes mental and conceptual tools as well as mechanistic tools.

When seen from this perspective, it becomes clear that the questions being asked by the researcher are the basis of good science and the tools used to answer the questions include anything that works. Radiesthesia undeniably works. So the sensitivity of a man, such as Turenne, lay in his ability to see that there were some unexplainable resonant angles picked up by his nervous system, angles belonging to no known element of his day. As is true in other disciplines, there were many radiesthesists who would not have bothered to question the existence of these angles but would have ignored them because they did not coincide with the accepted view of things. In this same way many who call themselves scientists today do not have the basic curiosity of the true investigator, but instead are mere gatherers of data to support an accepted view of reality. Such persons could have the most exquisitely sensitive instruments and not achieve the insights of a person with crude instruments who had the necessary inquisitiveness, the necessary questioning of all basic assumptions about the reality of things.

Since everything in nature radiates a fundamental frequency that corresponds with an angle of absorption, it is a simple process to go through all the angles to find out if there is a corresponding crystal someplace on the earth. In actual practice we usually find mixtures of different elements crystallizing together and so various resonance angles are found to correspond with samples being tested. As each element resonates with a specific color we find that these complex crystals are resonating to the various colors of the spectrum in varying proportions depending on their constituents. If we see the human bio-

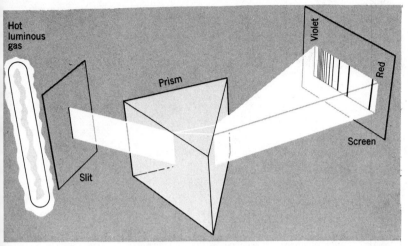
Line spectrum of a luminous gas.

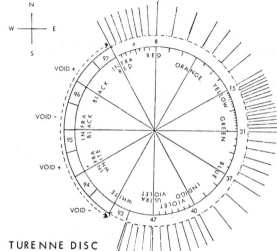
TURENNE DISC

structure as being composed of liquid crystals we can see that the proportions of the different colors involved in the pattern of crystallization are none other than the colors of our auras.

This is the same basic relationship which allows us to use spectrophotometers to analyze chemical composition. With spectrometry we must build a diffracting lens such as a prism or grating which incorporates the proper angles in order to separate the radiation from the sample into different bands. By vaporizing the sample through intense heat we can feed the resulting radiations through the diffracting lens and so get a photographable spectrum.

With our supersense we can skip all this work by using our consciousness to tune in directly to the very subtle radiations emanating from the sample and thus find the fundamental resonance angles. The results are similar, but supersensing, unlike spectrometry, doesn't require the elaborate technology or the total disruption of the life system (by burning it up) in order to study it.

BEYOND OUR SENSES

After years and years of programming ourselves with self-images of being physical bodies bound in time and restricted in space, it is hard to acknowledge our other dimensions of being. Thanks to our faculty of supersensing, however, we can experience ourselves as multi-dimensional beings free to soar in consciousness anywhere in the cosmos. Though our experiences will not be in terms of familiar physical sensations, they will be just as real. Once we learn to identify with our subtler perceptual faculties and start to use them, our self-image can expand beyond the limits of our physical bodies into the universal field of consciousness in which all vibratory forms and systems are interdependent participants in the universal hologram.

THE HEALER WITHIN

A powerful group of wholistic health techniques known as radionics are emerging in conjunction with the rediscovery of radiesthesia and ancient methods of knowing developed by the sages and prophets. The names Abrams, De La Warr and Hills, among others, give rise to images of black boxes with rows upon rows of dials and flashing lights broadcasting healing radiations into the wavefields of ailing patients. In supersensonics and radiesthesia we use our nervous system to detect subtle radiational fields and in radionics we introduce new radiational patterns into our vibratory wavefields to restore balance and health. The preferred method of supersensonic broadcasting of healing radiations over the past twenty years has been through sets of electrical circuits known as radionic broadcasters, but the latest bioenergetic advances in

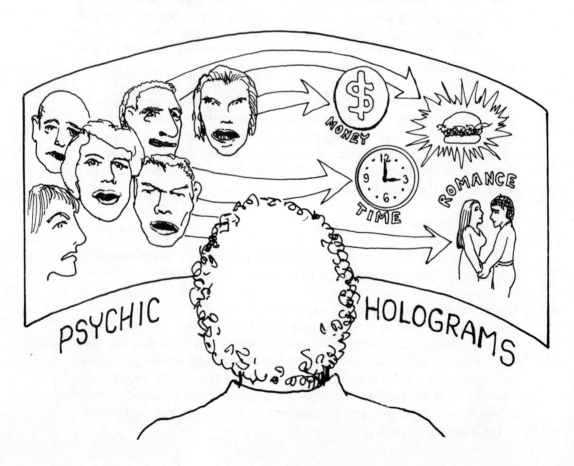

the field show that the healing radiations are actually broadcast by the real healer within us all, our consciousness.* The implications of this discovery are profound because we now know that the entirety of the reality we experience is a set of self-created circuits, a hologram through which the current of our consciousness flows. All our fulfillments and all our diseases are the results of such circuits. We have the power to choose how we want to experience our lives as well as the responsibility that such freedom entails. We all have the inborn power to create a vibrational field which is healthy. Even though many of us are not healthy at the moment, we can learn how to turn our minds into radionic broadcasters and introduce healing radiations into our consciousness. As in learning to ride a bicycle with the help of training wheels, the use of the first generation electrical radionic broadcasters is a good way to start.

OUR VIBRATORY SIGNATURES

Basic to the whole concept of radionic devices is the understanding that any object perceivable to our sensory equipment is essentially a bundle of energy vibrating to a unique pattern. Birds, trees, people, stones, water, planets, clouds, thoughts, emotions... any thing is actually energy vibrating to a particular pattern. How a pattern manifests physically as a solid, liquid, sound or color depends on the frequency and wavelength of its component vibrations. A stone, for instance, is energy vibrating at a slower frequency and longer wavelength than the gas hydrogen and is therefore more dense. This understanding has given rise to the notion that everything has a distinct vibrational "signature." You have a different vibratory rate than I do and our beings are each an expression of a special rate. Radionics has incorporated this fact into the language by referring to everything in terms of its rate.

Early researchers into medical applications for radiesthesia discovered that some sample of a human being such as a fingernail, saliva or hair, acted as a direct vibrational link with that person and that an accurate diagnosis of a patient could be carried out just by testing such a "witness." Because the fundamental rate underlying the crystallization of light into each person's body is at work in each cell of that body as a hologram, the vibrational pattern of all cells, from fingernail cell to brain cell, enable an operator to tune into the whole system's vibrational blueprint from a sample set of cells. These fundamental rates of vibration and pattern transcend our normal time/space constraints in a remarkable way. Merely by having a sample of cells as a witness, a trained dowser is able to find out what is happening to the rest of the body's cells even though the witness and the parent body are thousands of miles apart.

* See wave fields of the human hologram in "Supersensonics."

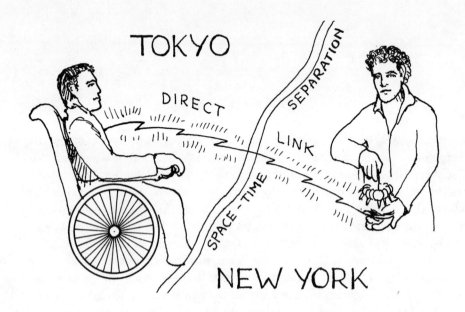

With a "witness" or sample of cells from a person in Tokyo, the supersensonics operator in New York is able to tune in directly to the man's vibration and find out what is happening to him in Tokyo.

A DIAGNOSTIC BREAKTHROUGH

These early investigators next discovered that the most effective remedy for a particular ailment could be found by dowsing over a series of sample remedies. For a person who had a bronchial infection they might take a lock of his hair as well as a bit of his saliva on a tissue as a witness in order to "tune" into his rate and then dowse over various possible remedies (medicinal herbs for instance) to find out which one would be most effective, what the optimum dosage would be, how often to take it, and when to discontinue the treatment.

Around the turn of the century the Catholic church was very enthusiastic about this application of radiesthesia because it greatly alleviated the need for trained physicians at their many missions throughout Asia, Africa, and South America. In fact the Abbé Mermet, one of the pioneering proponents of radiesthesia, was trained in these methods for later work in Brazilian missions.

The next development came with the insight that these dowsing doctors were dealing directly with the vibrational patterns underlying each illness rather than the superficial symptoms arising out of these patterns. Sadly, the magnitude of this insight has yet to be appreciated by the mainstream medical profession and over 90% of our present physicians rely on diagnostic techniques which are superficial in comparison. While the skill and precision surrounding modern surgery and pharmacology are praiseworthy in many respects, the simple truth is that to prescribe the proper cure we need to know as precisely as possible what it is we are curing. Modern medicine doesn't have diagnostic tools which are able to take into account each patient's unique psychophysical situation. In other words, when an ordinary general practitioner gives a checkup and afterwards prescribes some medication for inflamed tonsils, he is basing that prescription on the assumption that we are similar enough to be able to give all "normal" people the same drug when we exhibit similar symptoms, in this case inflamed tonsils. That assumption about humans being similar is valid only to a certain level of diagnosis, the level at which we all have similar organs and cell structures. When we begin to go into finer details, however, the uniqueness of each person's biology becomes more apparent. When we get into the really fine differentiations, at the level of the vibratory rates of molecules and atoms, then the only diagnostic tool that is an effective detector of atomic events "in vivo" is radiesthesic analysis. While it is possible to detect molecular vibrations with modern scientific tools, these tools do so at the expense of the molecules, which they have to disrupt to detect.

Rarely do modern pharmacological methods get to the roots of disease and when there is a "cure" it is usually the result of stimulating the patient's self-healing processes into hyperactivity. This hyperactivity in itself leads to future repercussions. An exa.nple of this is the use of corticoid drugs for rheumatism. These drugs stimulate the adrenal glands into hyperactivity which results in temporary relief from the symptoms of rheumatism but eventually the adrenals can no longer work at these accelerated rates and they malfunction. At this point the old symptoms often reappear stronger than before and the adrenal glands then malfunction as well.*

Many doctors who are also homeopaths have been quick to seize on radiesthesia methods and the fields of medical radiesthesia and radionics are rapidly gaining enthusiastic adherents among progressive physicians in Europe and the United States. Serious research has been delayed in the United States for many years because of the stringent laws requiring strict licensing for all medical practitioners. In effect this has meant that not even those who have been trained in medical schools sanctioned by the American Medical Association have had the freedom to use radionic methods for healing. Anyone practicing healing techniques without a license will be sent to jail. Even though there was a substantial body of evidence backing up the effectiveness of radionic methods, the FDA, prompted by the American Medical Association, chose at the outset to cry "quackery" and mounted an intensive public relations campaign to discredit these new innovations. In sharp contrast, the European

* Dr. Henry Bieler, "Food Is Your Best Medicine."

medical establishment did not adopt such narrow-minded attitudes and many professional associations sprang up throughout Europe for the express purpose of incorporating these effective new techniques into their medical practices.

HEALING VIBRATIONS

Dr. Albert Abrams, a San Francisco physician around the turn of the century, actually stumbled onto the discoveries which led to radionics even though he was not a radiesthesist. He was studying tubercular patients using the technique of percussion and discovered that they all gave off a certain dull sound when tapped on the abdomen when facing west, but not when facing any other direction. Intuiting that proper orientation somehow related this phenomenon with the earth's magnetic field, he began to investigate more closely.*

He next discovered that if he held a sample of tubercular tissue closely to a healthy person's neck and percussed his abdomen while he faced west, that the characteristic dull sound associated previously with tuberculosis would be heard. Upon moving the diseased tissue away, however, the sound could no longer be produced. Somehow the mere proximity of the diseased tissue was enough to affect the bioenergetics of a healthy person's wavefield.

His insight was that the vibratory pattern of the disease was being transmitted like a radio signal across space and received by the nervous system of the healthy subject. Abrams surmised that the subject was taking in the vibratory pattern of the diseased tissue, adding it to his normal pattern and thereby creating a new pattern which resonated with the signals from the tubercular tissue. This new pattern was symptomized in the dull thump which Abrams could hear when he tapped the subject's abdomen.

His next experiment was to run a wire from the sample to the patient in order to see if the patient would respond and the same dull sound was heard! Though Abrams knew little about electricity, he hit upon the principle of electrical resistance in searching for possible explanations for the success of transmission via a wire. In other words, the signal travelled along the path of least resistance, in this instance the wire, like water seeking the lowest level.

---

* Modern radiesthesia confirms this relationship. It has been found that facing south or east is invigorating, facing north is calming and facing west is debilitating. This was common knowledge among Chinese, Egyptian and other early agrarian societies including the native American cultures. We can easily test this out for ourselves by sleeping with our heads pointing in these various directions on different nights and comparing the experiences.

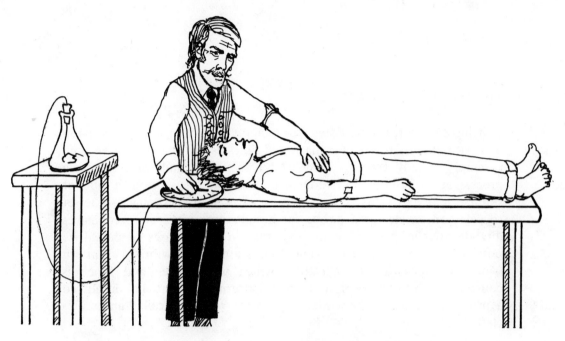

Dr. Abrams had discovered how to diagnose diseases from their vibratory emanations.

At this point in his experiments he noticed a sound different to that previously associated with tubercular tissues and discovered a jar of cancerous tissue close to the patient. Upon removal of the jar from the immediate area the sound disappeared. This was a major discovery because it suggested that different diseases could be diagnosed via the differing reactions they elicited in a person's nervous system.

The main task now was to figure out a more accurate method of distinguishing these responses from one another, since identification via differing abdominal sounds was too crude. Abrams had a sudden intuition: why not interpose some measure of resistance like a rheostat in the wire running from disease sample to patient and find out which setting resulted in an abdominal sound?

This setup worked admirably and quite soon Abrams had found a number of specific settings for disorders such as cancer, tuberculosis, malaria, and strep throat. Abrams had discovered that disease could be dealt with as a vibratory phenomenon and therefore simplified the whole business of proper diagnosis. The key was the human nervous system; the abdominal reflexes of his patient had proven that the human nervous system was capable of identifying diseases via their subtle vibratory patterns. Even though he had no background in radiesthesia, Abrams had reached the fundamental understanding that the human nervous system is an extremely sensitive receiver, more so than any electronic device, constantly taking in all the different vibrations radiating in from the universe and somehow distinguishing them from each other.

THE FIRST BROADCASTER

Being a doctor, his next step was to apply his findings to the question of what remedies to give for diagnosed illnesses. By running a wire from a sample remedy through a rheostat to a person and using the same process of percussing the abdomen he was able to work out a table of suitable remedies. This was still not what he wanted though, so he set to work on an electrical device for broadcasting vibratory frequencies directly to the patient. Reasoning that in many cases it should be possible to replace actual chemical ingestion with a vibratory pattern (since the physical crystallization which we can see and touch is constructed to the tune of a vibratory pattern) he eventually developed a prototype radiational broadcasting device which he called an oscilloclast.

In its simplest form the oscilloclast was a circuit consisting of a lead running from a battery through a rheostat to an electrode placed over the patient. Through this circuit a vibratory frequency would be "fed" into the wavefield of the patient to alter the disease-causing vibrational rate. When properly tuned with the rheostat the new frequency would balance out the unhealthy rate and create a wholesome vibratory rate in the patient. Once this healthy rate was established, the patient's self-healing processes would begin functioning normally and eliminate the dangerous toxins from his system.

In diagnosis, where rates for a particular disease and its remedies were sought, the sources of the vibratory rates were the samples being tested. In the earlier example of cancerous tissue affecting the nervous system of Abrams' patient, the diseased tissue itself was emitting a minute radiational current. Though the strength of such a current was indescribably small, it was still strong enough to be detected by the human nervous system. In treatment however, where a tuned current that would counteract the vibrational field of a disease was required, Abrams started off by using a battery as the source of the original current which would later be transformed by the circuit into the healing vibrations.

Radionics therapy was started through this work by Abrams. His work was a practical example of effective therapy using electro-vibratory fields instead of the surgeon's knife or massive chemical suppression of symptoms. It was a practical medical application of the new findings in physics (in the early 1900's): that all matter was made of vibrations. This was as radical a breakthrough as the germ theory of disease which caused diagnosis and therapy to make an exponential leap in the late nineteenth century because, like germs, the vibratory wavefields of matter precede observable physical symptoms. Most importantly however, these fields precede germs because disease germs only live in tissues that have already undergone degeneration of some sort and degeneration starts with a change in the vibratory "life" field from which the cellular tissues take their form.

The next forty years of work in the field of radionics led to many modifications in Abrams' original design of the circuitry in radionic instruments, but no changes in the basic theory. Most of the modifications consisted of adding more tuning dials for more exacting differentiations. The basic idea behind the instruments still consisted of running current through a coil and changing the length of the coil in order to tune in to different frequencies. However, De La Warr found that electrical current was unnecessary and that something else was responsible for transmission of the biophysical signals.

The circuits used in radionic therapy were likened to those necessary for a radio. The patient's body or the specimen being tested (mineral or plant, for instance) was the transmitting station sending out its signal patterns. The wires, coils and rheostats acted as the tuner. The nervous system of the person receiving the signal through the wire acted as both the antenna picking up the signal and as the amplifier which boosts the signal. And the neuro-muscular system of the receiver acted as a speaker to make the signal of perception sensible to our senses.

A patient transmitted a signal through the wire and coil to the receiver. The operator, who could be either the receiver himself or someone monitoring the nervous system of the receiver, changed the length of a coil in the wire using a rheostat until the coil's length was in tune with the signal coming from the specimen. The operator knew when the resonant tuning had been found by an obvious reaction in the nervous system of the receiver, such as a dull thump when his or her abdomen was percussed or a "stick" occurred when rubbing the patient's skin due to the operator's skin resistance.

Dead tissues, when diagnosed through a rheostat, would register a reflex in the abdominal muscles of the test subject, but when passed through a second rheostat in series would register nothing. Living tissue, on the other hand, registered even stronger reflexes when passed through a second rheostat. This meant that the circuit was not operating on the basis of electrical resistance because in that case the reflex would be weaker at each successive rheostat. Some other type of energy was at work but no one could say what it was. At this state of development the exciting factor was that the circuit worked.

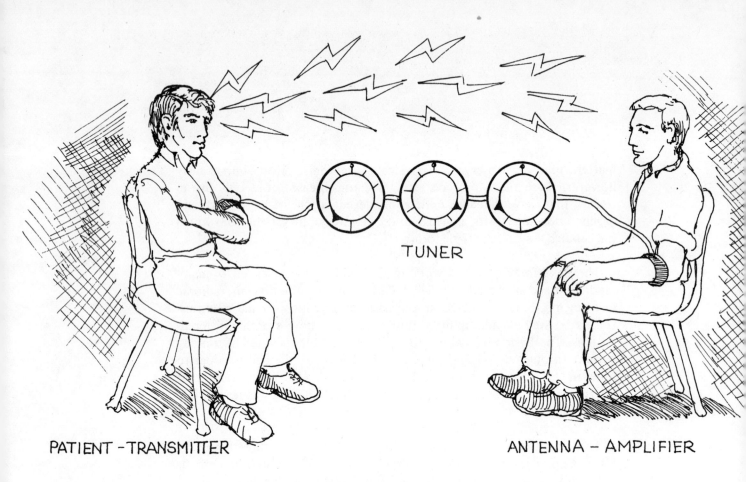

PATIENT - TRANSMITTER   TUNER   ANTENNA - AMPLIFIER

Each rheostat served to create a specially tuned resonant antenna which filtered out extraneous vibrations and only passed on the resonant ones. As this process continued through successive antennae or rheostats the signal became increasingly refined and the strength of the subject's reactions was a measure of how refined the incoming signal was. By stringing several of these coils in a series so that the current had to pass through several of them, these researchers could increase their exactness. If the first coil was used to tune in to different frequencies on the order of thousands of Hertz for instance, then a second coil could be added to the wire which would further tune the frequency on the order of hundreds of Hertz and a third coil for tens of Hertz and so on.

The subtle sensitivity at work in tuning these circuits is also found in homeopathic medicine where the strongest or most "potentized" remedies are the ones which contain the finest quantity of active ingredient. Somehow our nervous system reacts more strongly to highly refined quantities than to massive quantities and the level of refinement which can be detected by our nervous system is staggering. As an example consider that the strongest homeopathic remedies are the 10,000X potencies. If we take an undiluted solution of remedy and dilute it one part in ten we have a 1X potency. If we then take this 1X potency and dilute it again one part in ten we have a 2X potency. So 10,000X is so diluted that even spectrometric analysis or electron microscopy cannot detect any of the original substance. Yet these potencies have strong effects on the user! That is how sensitive we are.

One of the authors with an Abrams type Diagnostic and Treatment Unit. The wire on the left leads from the witness receptacle with its multi-colored amplifying lights to the circuit board. The round disc he is stroking to the right is the stick pad.

**Part of collection
-- MUSEUM OF ANTIQUE DIVINING INSTRUMENTS**

THE STICK PAD

Later research revealed that the operator could also be the receiver by using a device known as a "stick pad". Operating on the principles of galvanic skin response, the stick pad was a small plate made of slate or smooth wood covered with a thin rubber membrane attached via a wire to the radionic circuit. The operator stroked it very lightly with his fingers and when a resonance occurred between the transmitted signal and the coil-rheostat an excited signal was conveyed to the nervous system of the person stroking the stick pad. His nervous system would detect this excited signal and in turn respond by causing the static charge in the skin on his fingers to change. Instead of gliding smoothly across the pad his fingers would stick slightly to the membrane, which would pucker up. Abrams had an electric circuit beneath the stick pads on his later units in order to give them a tiny static charge which would enhance the puckering effect.

COLOR HEALING

The unit pictured above was the result of several modifications to the first oscilloclasts and diagnostic units developed by Abrams. Both units were combined into one instrument. The round box with seven colored lights (red, orange, yellow, green, blue, indigo, and violet), was designed to hold either a witness of the patient or a wire from him to the instrument, in order to irradiate the incoming signal pattern with its resonant colors and thereby amplify the signal.

All matter is actually a crystallization of light resulting from absorption and concentration. Each angle of absorption will be in one of the bands of primary color -- red, orange, yellow, green, blue, indigo, or violet -- and will resonate with any manifestation of that color. To prove this we can take any bit of matter and turn it back into light by vaporizing it and then shooting that light through a diffracting prism onto a screen. We will see certain colors projected onto the screen in varying proportions. These proportions of only certain colors are the color autograph of that bit of matter and form the foundation of the scientific technique known as spectrometry.

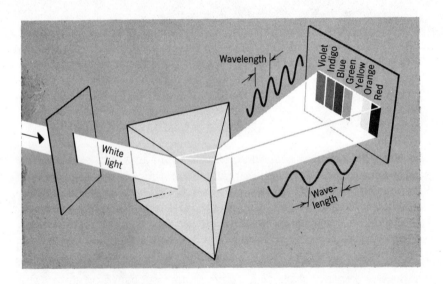

While each person resonates most strongly with one or two of the colors we are all made of seven vibratory patterns of subtle energy called chakras. These are fully described in the theory of nuclear evolution. The theory behind color therapy in nuclear evolution is that deficiencies can be corrected by irradiation with the resonant color of any deficiency. For example, if we are suffering from sluggish intestines we might irradiate the intestinal area with red-orange light, or do color breathing exercises with red-orange, or use an aura pendulum set to the orange vibration. The color in itself does nothing, the active ingredient is the resonant consciousness of the patient. All the color exercises do is stimulate that area of our being corresponding to the particular color we are concentrating on. This stimulation helps us heal that deficient facet of our self.

TUNING IN

The radionics set pictured above is representative of the basic design principles. A stick pad was used for identifying resonant tuning. The tuning dials of the unit were organized into two sections, a "personal" section and a "treatment" section. The left hand rows of dials are for the "personal" rate, that of the patient's disorder, and the right hand rows are for setting the "treatment" rate, the vibratory pattern added to the "personal" rate in order to heal it. The witness of a patient was placed in the colored light receptacle which in turn was connected to the stick pad via the tuning dials. The idea behind the rows of dials was fine tuning. As researchers began to discover the rates for different bacteria, they found that two different bacteria had the same rate on the simple one rheostat tuner. The next step therefore was to add a second rheostat in order to fine tune the signal coming from the first. For instance, if two bacteria registered the same reading of six on the primary dial then the second dial would differentiate a rate of sixty-two from a rate of sixty-three (each of these dials being marked from one through ten). Eventually researchers expanded up to five rheostats in order to be able to account for all the finer differentiations. The next jump came when they started to add more than one row of dials in order to tune for what they referred to as the "subtle bodies" underlying the physical. Each researcher usually had some background in metaphysical belief-systems and so each one tended to lend credence to models of being which included subtler "bodies" or energy fields such as an emotional "body," desire "body," mental "body," psychic "body," and spiritual "body," the labels depending on whether they were Theosophists, Buddhists, Rosicrucians, Hermeticists, Druids, etc. Each of the "bodies" had a subtle rate all its own, they believed, and they found their results to be more effective if they set up a rate for each of them. Soon their instruments were being designed with rows and rows of dials. Abrams himself included four rows: one for the "personal" or basic physiological rate, another for "intensification" of the personal, a third for fine-tuning into specific organs (he called it the "visceral" rate), and a fourth for the mental pattern behind the grosser physical symptoms which he called the "identification" rate.*

Actually many investigators found that as a rough rule of thumb the "treatment" is usually one setting less than the "personal" rate; in other words, each dial is set to one number less than its counterpart among the "personal" rate dials. As numbers represent natural forces, an odd number next to an even number brings in the opposite force to interact

---

* Those interested in investigating the specifics are encouraged to read "The Chain of Life" by F. Guyon Richards and "Medical Radiesthesia" by Vernon D. Wethered.

with the force represented by the even number.* Many radiesthesists did not bother to find exact "treatment" rates, the notable exceptions being researchers like Abrams and De La Warr, who designed special "treatment" components into their units.

RADIONICS WORKS WITHOUT ELECTRICAL CIRCUITS

The next big breakthrough came with the work of Englishman George De La Warr. One evening in his lab he tuned his unit to a treatment rate but accidently forgot to plug it in to the current. The next morning he found out that the patient had been treated anyway! So why bother building complex electrical circuits with rows of rheostats and wire coils? All that was really needed was a dial that turned inside a circular "antenna" made of a piece of copper wire.

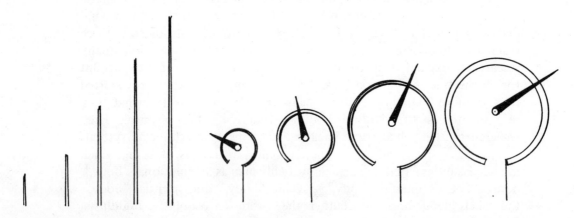

The dial on the face of the radionic broadcaster is really a tuner for a circular antenna inside. The knob slides a curser around the circumference of the antenna until the length resonates with the operator's thoughts and there is a "stick".

Christopher Hills, at one time a co-worker of the De La Warr family, has described this situation as very similar to the tuning of a radio set, which does not need current flowing through it to be tuned to the proper signal. The only process that current is required for is to drive the amplifier, which boosts the signal through speakers which in turn vibrate at rates which our nervous system detects as audible sounds. Any length of conductor, straight or coiled, is an antenna with proportional resonance points on it. The only difference with varying lengths is that

---

* For a deeper discussion of the relationships between numbers and natural forces and how numbers act as witnesses for subtle natural dynamics, see Volume III "Supersensonics" and the forthcoming Volume IV of "The Supersensitive Life of Man" series: "From Zero to Infinity -- the Divine Mathematics of Proportions," both by Christopher Hills.

the space between these resonance points is longer on a longer antenna and allows for a finer discrimination of subtle differences, (See "Supersensonics" by Christopher Hills for details.) By touching the antenna at different points along the length with the end of a tuning dial we change the frequency with which the antenna resonates. This is exactly the same principle involved in the playing of a string instrument. As the musician moves his finger to different spots on the string frets and touches them, different frequencies or notes are produced. Tuning involves finding the length of antenna which produces a frequency in resonance with the bioenergetic signal produced by the patient.

De La Warr continued to build complicated wiring into his sets so that his customers would see a complicated piece of circuitry and feel better about what they were getting for their money. Unfortunately one of his customers, a lady who said she felt cheated out of her one hundred pounds when she saw what she got for it, entrapped him by going to him for treatment and proclaiming him a fraud afterwards. There was some question of the lady "setting up" De La Warr since a number of medical scientists regarded the whole field as pseudo-science. "The Black Box Case" was quite a famous trial as many orthodox physicians took this opportunity to proclaim radiesthesia and radionics "quackery." Though De La Warr won the case and got a farthing damages against the lady who had sought "legal aid" under the state, the trial cost him many thousands of pounds and almost ruined him. He eventually had to rely on donations to continue his research but fortunately found many backers.

One of his backers was Christopher Hills, for whom De La Warr constructed an unusual instrument which did not need rows of dials because it was pre-set. Hills had realized that some rates are constant, for instance the Cosmic Intelligence which is behind and in everything in the universe helping it to become itself. They set out to design a device which could be pre-set to the rate for Cosmic Intelligence. The idea was to mass produce a device for a few dollars which could fit into anyone's pocket, a device constantly tuned to the Cosmic Intelligence. They simplified all the circuitry into a simple little box so anyone could walk around with Cosmic Intelligence but Hills eventually abandoned the project because he "realized that if you make Cosmic Intelligence only worth a few dollars then nobody would want it." He also felt the universe had its own way of dealing out cosmic intelligence when people were ready and so cancelled production plans because the principle of the instrument could have been misused for psychotronic warfare and psychic imprinting of holographic patterns of thought on other people.

THE REAL CIRCUIT IS IN OUR CONSCIOUSNESS

It was about this time (1960) that another major breakthrough in

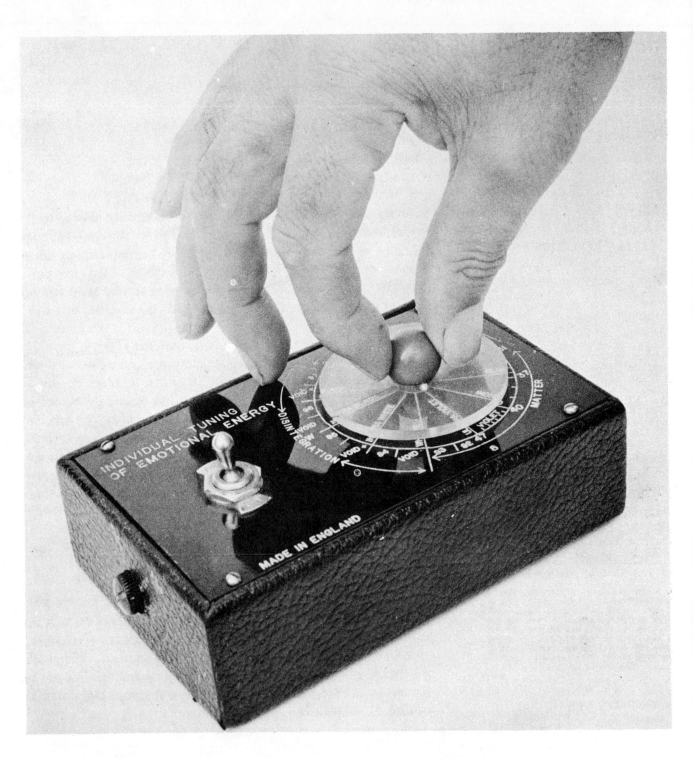

This Supersonic kundalini device detects a person's aura color and determines which of 44 natural elements is predominant in the physical body. Inside the unit there is a whole octave oscillator tunable from 200 cps. to 400 cps. as a physical exciting stimulus while the harmonics of the radiomagnetic field of the horizontal wave field of the emotional body are tuned by a magnetic dial at the center of a Turenne disc. The "thought fields" of the vertical wave field are patterned by an array of prefixed De La Warr permanently tuned dials set at the rate for higher creative intelligence in the east-west wave field. The set was designed by Christopher Hills and custom made to his specifications by George De La Warr of Oxford, England in 1960.

radiesthesia and radionics took place through the work of Hills, a successful businessman turned yogi and consciousness researcher. Because a patient was treated while the unit was not plugged in, De La Warr had assumed that the machine did the work by itself. Hills disagreed, saying that because De La Warr had created the circuits in his mind before manifesting them in physical bits of copper and plastic, what really switched them on was the hologram of his own consciousness thinking about them being on. In other words, the machine was only an extension of the circuits in De La Warr's mind, a way of making them physically real. All the electrical paraphernalia was not doing the work, the dial pattern "rate" just focused the consciousness of the operator into a cosmic hologram.

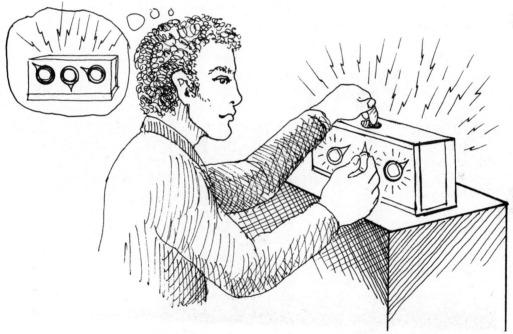

"The healing circuits are created in the operator's consciousness."

CONSCIOUSNESS MAKES THE CIRCUITS WORK

Previous French research in radiesthesia supported Hills' insight. When Louis Turenne discovered the idea of a "resonant antenna" to explain the action of a dowsing rod he had first used a length of metal rod. He soon found, however, that any straight length, even a line drawn on paper, would operate as an antenna. Testing this finding with magnets and with other radiesthesists he soon determined that a two-dimensional symbol of a three-dimensional form worked just as well to focus the consciousness of the operator into specific patterns.

The beauty of having a piece of equipment is we do not have to figure it all out in our mind. It enables us to attribute numbers or other proportional values to particular vibrational patterns and have those values concentrated in a material representation. This saves us the energy needed to get the same result by purely mental concentration and frees the mind for other work. In other words, the instrument with its various settings acts as a wave guide for the consciousness of the operator, whose belief that the instrument does something makes that something happen. All the gadgetry really does is help the operator focus and concentrate consciousness into a certain pattern which acts like a hologram for cosmic intelligence to work through.

Though many who take pride in the accomplishments of empirical science and have a vested interest in the western conception of reality will probably dispute this immediately, the simple truth is that the stuff we use to think with, consciousness, is indistinguishable from anything we think about. Everything we experience as an external world beyond our skin is actually a mental representation within our consciousness, so it is scientifically impossible to separate the two -- the knower is not separate from the known. The next frontier for science then is mapping the circuits of consciousness, because until we can understand the nature of the stuff we are using to experience life with, we can never be sure that what we experience is not a delusion in our consciousness.*

BELIEF SYSTEMS

The primary circuits through which we express our consciousness are our belief systems. A belief system is the set of assumptions and values which mold our perceptions of the total environment around us. Each one of us has a unique way of perceiving which is based on a sense of conviction about basic assumptions that we believe are true but usually cannot be verified, such as our assumptions about time. The same holds true for societies. Each society has its own set of assumptions about the nature of disease and health, and consequently a special healing technology. Yet there are many different paths which all lead to health. The essence each culture shares is a sense of conviction that their methods lead to health; they have proven beyond any doubts that their techniques work. How do they prove it? Let's take the example of Pasteur and the germ theory of disease.

*"Supersensonics," by C. Hills

When Pasteur proposed that illness was caused by external agents called bacteria he was ridiculed, yet when he showed that introducing certain germs into any person resulted in tuberculosis he was taken seriously. In order to make believers out of his critics he had to show them a process which they could sense with their own senses and repeat themselves. Once they had seen bacterial infection with their eyes and repeated the process with their hands they had experienced germ theory physically for themselves and took it to be real. They believed that what they could experience through their senses was real and what they could not experience through their senses was not real. One consequence of having such a belief system is that any proposed theory of disease which does not include sense-perceivable material agents as the cause will not be taken seriously.

Leaps in understanding start from accepted beliefs and stretch into unknown areas. Abrams thought of his discoveries in terms of wires, radio propagation and resistance because those were current concepts which could begin to explain the phenomena. Without some conceptual foundations to which new discoveries can be related there would be no way of describing the discoveries. In order to describe phenomena linguistically, we have to use words which are already referring to a past experience and already associated with certain images in our minds.

Abrams was effective with wires, electrical circuits and electricity, so he believed they were necessary for effectiveness. De La Warr found he could be effective without using electrical current and without wiring together an electrical circuit, so he believed that a healing energy could be channelled through a series of antennae. Hills found that no apparatus was needed and the real circuits were mental – in consciousness itself. While tuning the dials does help to focus consciousness they are not necessary. Each of these refinements of radionic broadcasting showed previous assumptions about essential prerequisites were only necessary to make the theory fit into the operator's particular belief system.

BELIEVING IS HEALING

The key to health is belief. Belief activates the circuits we create in our minds and our sense of conviction, whether true or false, sets our consciousness into the vibratory patterns we experience as life. For centuries we have known this, and many groups have taught their followers that if they believe strongly enough their faith can prevent and cure diseases. In this sense our western world view is similar to the views of the shaman and mystic, all are models of the actual which their adherents believe in utterly. Western science's wonderfully detailed preciseness regarding the physically manifested world makes it a brand of magic possible for a larger number of people, and this is its power. Looking at the art of healing through comparative examples can help us see a common essence underlying different techniques. By taking a brief look at what a shaman does, what an M.D. does, and what a radionic therapist does, we can see the underlying essence.

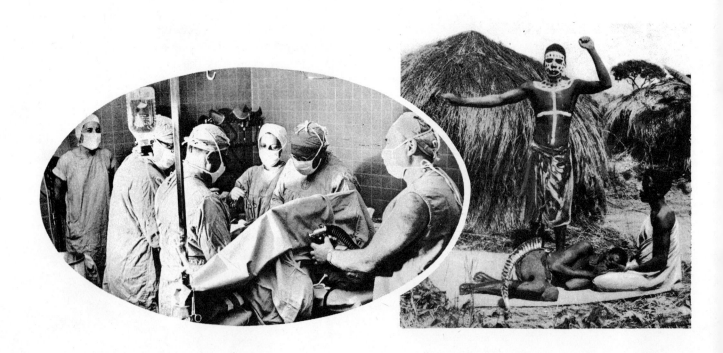

A man gets swollen tonsils in the jungles of Darien in Panama so he visits the local "curandero" (spanish for "healer") and asks for help. The curandero gives him some herbal teas and performs a ritual in trance wherein he invokes "the powers" to help his friend, whom he advises to rest for a while.

Another man gets swollen tonsils in San Francisco and visits the local clinic for help. The resident M.D. takes some tests such as a throat smear and then gives him a prescription for some antibiotics. He advises his patient to get plenty of rest.

A third person gets swollen tonsils in London and so he visits a local radionic therapist (radiesthesic treatment is legal in England) who finds the identification rate for the tonsils and then sets up the treatment rate. He also advises plenty of rest so the body can heal itself.

The curandero's ritual is actually an exercise which focuses his consciousness into a pattern corresponding to a healthy patient through powerful visualizations coupled with herbal medicines. The curandero acts as a channel for transmission of a vibrational field which can temporarily balance out the disease-causing pattern within the sick man's being. He knows that deeper cures must be used in conjunction with the treatment of the physical symptoms via his stock of plant medicines. If the sick man is to be helped at all he will also question the man about his living habits and suggest appropriate changes. The primary tool of the curandero is an intuitive insight into the nature of the sick man's problem which feels what is wrong, as if the curandero were inside the sick man (this faculty is associated with the indigo or sixth chakra).

The medical doctor's diagnosis on the other hand is based primarily on a large body of statistical evidence correlating physiological symptoms with pharmacological remedies which will suppress the symptoms. The M.D. will try to view his patient objectively in terms of his symptoms and for this his main tool will be his intellectual and conceptual faculties (the yellow and blue or third and fifth chakras). The curandero will intuitively feel his way into the problem as if it were in him. The M.D. will intellectually analyze the symptoms as an external observer, correlate them with the corresponding definitions in the pathological index and prescribe drugs from a pharmacological index such as the Merck Manual. These drugs will generate powerful vibratory fields within the patient's wavefield which, like the curandero's thought waves, will counterbalance the disease-causing vibrations. However, because these drugs are symptom suppressants, they seldom affect the inner roots of disease and relapses are frequent.

Such inner roots are deep within the patient's being and long before physiological symptoms appear, the radionics operator can detect changes in his patient's vibratory rate and then broadcast balancing vibrational waves into the sick man's wave field. By viewing all material phenomena in terms of underlying vibratory fields, the radionics operator makes use of the best in both approaches. Via the scientific precision of the radionics dials he is able to intuitively tune in to the vibratory imbalance in his patient and determine the exact vibratory wave field which will restore balance. The advantage of his dials over the curandero's visualization is that the dials will keep on acting as a waveguide to channel the healing vibrational pattern even when the operator has switched his attention to a new situation, whereas the curandero's channeling requires constant attention.

The curandero, the medical doctor and the radionic therapist all have found methods which help the patients rebalance their vibratory rates and heal themselves. If either the patient or the healer doubts the methods being used, the effectiveness of these methods as tools for redirecting the consciousness of the patient into self healing patterns is automatically impaired because a counter-pattern of maintaining the illness is being programmed into the patient's vibratory field.

BELIEF SYSTEMS AND THE LEVELS OF CONSCIOUSNESS

The different therapies reflect different levels of consciousness at work creating a representational system of reality with which to focus consciousness. The curandero is the archetypal shaman who operates on the violet level with magic that the patient does not understand. The M.D. is the archetypal analyst who operates on the yellow level with analytically precise differentiations of the whole into biochemical components which he then augments or dilutes in order to maintain a balance. The radiesthesist is the archetypal blue level person conceptualizing (*con* = to gather, plus *cept* = knowing) everything as part of an all-inclusive vibrational field and using vibrations directly for treatment.

They each have their own types of gadgetry which they use in varying degrees to aid them in focusing their consciousness. Some need the gadgets to be effective because they cannot do it all inside their minds. That is, they cannot create the mental circuits that connect the future healthy pattern with the present ill pattern via specifically imagined transformations of the existing mental hologram thought patterns that are creating the disease.

This is an intriguing train of thought as it tempts us with visions of some innate power we have to shape our own reality. The ancients referred to this as the power of worship. *Worth* = having value as in "worth investing in" + *Ship* = to shape. Literally then, worship is how we shape our value system, how we shape our value for something enough so that it is worth investing our consciousness in it. Each of the three sick men is suffering the consequences of investing his consciousness in something which was not in tune with his inner needs. Whether it was a bad diet, overwork, smoking or negative thinking, etc. is not so important as the fact that they made choices which resulted in their present situation, choices which could have led to healthier states of being if they were identified with their real inner *needs* rather than their unnecessary *wants*.

Such choices are not easy for highly social animals such as humans because the values we identify with are often the values of the rest of society. This is the tyranny of "normalcy" where anyone who questions the values of the status quo is immediately suspect as a disturber of the peace and shunned by those of his fellows who are meekly content to be in comfortable ruts and will trade material security for their souls. In such a society much value is given to gratification of our wants because it is good for the economy; if we restricted our wants to our needs the economy which makes our comfortable way of life possible would collapse from a lack of sales. The outlook for health is not totally bleak though, because we have the choice to invest our consciousness in beliefs and values which will lead to inner fulfillment and hence to health.

Over the years countless belief systems have risen to change the world condition and a common denominator in all of them is the conviction and faith of their authors in some force or intelligence or god greater than their own minds, values and concepts. All have had a basic humility before the awesome cosmic force which has patterned basic components of the universe such as light and hydrogen into the fantastic orderliness of this planet our bodies inhabit. Whereas the originators who blaze a trail have to seek guidance from some intelligence beyond their understanding, their followers are able to ride on second hand models. Usually the originators are much in awe of the reality beyond them while disciples tend to lose sight of this original wonder and curiosity, and instead go for the security trip of preaching a dogma.

For example, many researchers into radionics and radiesthesia are not interested in opening up a direct hot-line to the cosmos because they want to tune in to valuable minerals or lost objects or heal tissues. Even though this technology is a key to discovering our purpose and being healthy, many researchers simply do not care to use these tools to investigate the basic premises underlying their belief systems. This is human nature, even if we are miserable the last thing we want to change are the thought patterns which are causing us to perceive the cosmos in a way that makes us miserable.*

THE HEALER WITHIN

For centuries our theories of health have placed responsibility on external agents such as spirits, demons, witches, gods, germs, and the state of society. Now the time has come for a complete about face. We now know that all those external agents are actually holographic images within our consciousness and we have the choice to be controlled by them or to control them. The paths of science and mysticism are merging and the way is becoming clear. Spirit and matter are interdependent parts of the unified field of our consciousness, the new frontier of health research. Theoretically we can prove that everything is *one* and not separated, but to actually *experience* wholeness or health is going to take major changes in our belief systems.

Such changes demand that we understand how the circuitry in our consciousness is wired, how the patterns in our mind create our experience of life. We have the power to choose the holographic patterns operating in our field of consciousness yet many of us are unaware of

---

* This whole problem is explored in the correspondence course, "Into Meditation Now" by Christopher Hills.

our choices and experience ourselves as limited to unpleasant circumstances. To awaken the healer within us, the first step is acknowledging the areas of our life experience in which we feel we have no choices; as long as we hide from our problems we will be controlled by them. By bringing them out into the open we can begin to question our problems and do something constructive about them.

The second step is to discover why we have the problem. Usually when we feel stuck in a painful mess we are also attached to some behavior pattern which is perpetuating the mess. Taking a sticky relationship with another person as an example, if we are so attached to someone that we feel we cannot break up the relationship even though we are unfulfilled, then we are asking for pain. By looking at our diseases as the effects of choices we have made in the past we can start to control our health. If we blame our problems on external agents we abdicate control because we have no power over another being's action. We can only change ourselves. Once we take responsibility we can often trace diseases or pain back to patterns of thinking and behaving which we can then decide to keep or throw away. If we decide to keep the pattern then we know who's to blame if we feel rotten.

The third step is discarding our disease-causing patterns and replacing them with healing patterns. By the time we have recognized how we are causing our problems we also know how to change our situation. Usually there is a little voice or feeling inside that lets us know what we have to do and we have the choice to keep our problem or throw it away. Christopher Hills has called this the *choiceless choice* because there is actually only one way to go. For instance, if we choose to keep a pattern that our inner voice says is unhealthy, such as procrastinating, and then rationalize it by saying that we are humble and really not interested in achieving anything, there will eventually be a day in the future when we look back at our life and realize with a shock that we have nothing to show for it. Then we feel really sick inside and wish we had listened to our inner being. Radionic treatment of diseases will only yield temporary results unless we also confront our fundamental patterns of sickness and change. Unless we actually change a negative pattern, a sickness will eventually return after a healing. The healer within us is quietly waiting to show us how to change and choose the healthy way of life.

Before we can follow our inner voice we have to hear it, and honest feedback from friends offers some of the best reflections of how well we listen. A group of our friends can often penetrate more deeply into the nature of our problems than we can ourselves because they can spot disease-causing patterns that we are attached to and do not want to look at.* Inner conflict is the root of disease and any health therapy that alleviates symptoms will only be temporarily effective if our inner tangles are not straightened out. The next step is to look at some practical experiments and supersensonic tools which we can use to find out for ourselves what new dimensions of health patterns, radionics and energy broadcasting can lead us to.

RADIONIC BROADCASTING

If possible find a radionic broadcasting unit, perhaps even one of Abrams' oscilloclasts. Though they are rare there are still a few around,

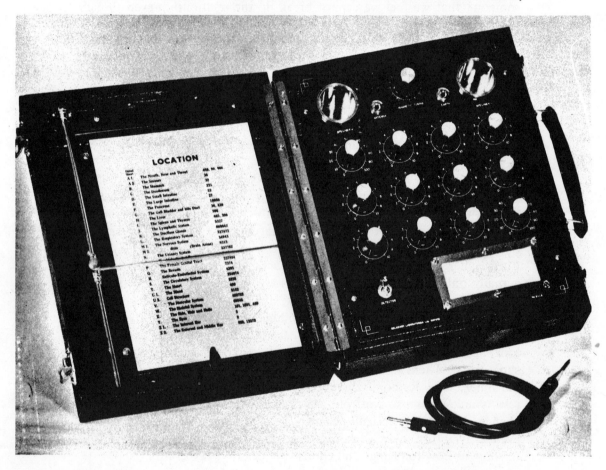

Photograph of the standard diagnostic instrument used by the De La Warr researchers.

* The methods of "Creative Conflict" outlined in Tape No. 16 of the Rumf Roomph Yoga series (available from the publishers) are designed for developing a group that can give us dynamic feedback. See also "Exploring Inner Space" and "Universal Government By Nature's Laws", both by Christopher Hills.

often in an attic or at a flea market. The process of using these instruments is quite simple, especially if it is a model combining diagnosis and treatment apparatus into one unit.

On the model pictured here, the personal rate is set up on the left hand set of rheostats and then the treatment rate is set up on the right hand sets. The key is the stick pad. Unless the operator's sensitivity is accurately developed the likelihood of error is high. This is why beginners are encouraged to double check themselves with a pendulum.

Plants serve as excellent test subjects. Take a leaf from one of two similar plants, determine its present rate and set it up on the left hand or "identification" dials. Next discover the rate for perfect health and set it up on the right hand "treatment" side of the unit. Note any changes in the test plant as compared to the control.

Next try this same experiment with a sick plant and see if it improves. If the results are not towards improvement it may mean that there is a more specific problem that needs attention such as infestation by aphids. Then the focus of the treatment would be on aphid control. To eliminate the aphids you broadcast a vibration that is opposite to the life "rate" of the aphids. The unit is only an extension of the researcher, whose own questioning and persistence lead to success.

FAIL-SAFE BROADCASTERS

While the range of rates possible on a classical radionics unit makes it quite versatile it is possible to achieve comparable results with so-called "pre-set" devices. In essence these devices are already tuned to those cosmic energy patterns which are subtly urging us to actualize our potentials. The Square Balance Equalizer for instance, is pre-set to the patterns of BALANCE and POSITIVE GREEN so that the responses we get with a Square Balance Equalizer are automatically in tune with basic natural forces. Such pre-set devices act as fail-safe tuners and reduce the number of variables we have to deal with.

One of the big problems in radiesthesia and radionics is to get the subtle subconscious desires out of the mind or else thay affect the accuracy of any finding. Over many years of testing it has become clear that whatever motivations or desires are uppermost in our mind while divining are mirrored in the findings. If these motives are for selfish gain the divining faculty will not be accurate. Selfishness implies a separatist attitude that wants good for itself with little concern for the other members of the total enviornment. Health is a process of interacting dynamically with all the other parts of the cosmic hologram so that we are motivated by concern for the welfare of the whole rather than one piece of it. The divining faculty operates most effectively when the operator is concentrating so intensely on the vibratory pattern he or she is seeking that no thoughts of the separated self enter into consciousness. Pre-set devices such as the Square Balance Equalizer, Life Force Charger, Magnetron, Broadcasting Grid and Pi-Ray Orgone Accumulator Coffer of supersensonics work quite effectively as broadcasters and so belong to a new generation of radionic instruments which combine the best features of the electrical circuit models with new discoveries about psychotronic generators. This combination allows a remedy's vibratory pattern to ride piggy-back on forces of Nature which are already involved in our being what we are, and so integrate from within our energy system rather than disrupt it from without.

THE SQUARE BALANCE EQUALIZER

Thousands of years ago the concept of balance was introduced as a means of measuring the proportional value of trade goods in the market place in terms of relative weight. In turn, modern physics and supersensonics have discovered the importance of the null point where opposites are reconciled and balanced. In radiesthesia and supersensonics this balance point deals with very subtle energetic balances, for instance the balancing of the sympathetic and parasympathetic nervous systems or the balancing of extensor and reflexor muscles. Every time we have any nervous reaction -- even muscular reactions -- there is an electrochemical balance at work involving the secretion of chemicals such as choline and acetylcholine which play a pivotal role in the muscular movements. In fact the neuro-muscular reactions which we see amplified by diviner's rods and pendulums are nothing more than the results of chemical processes which in turn are triggered by neural responses to incoming radiations.

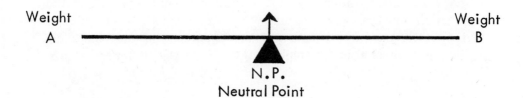

Early radiesthesists and diviners discovered a method for determining whether a remedy was right for a patient by utilizing the ancient principle of balance. By putting a witness of the patient, such as a lock of hair, on one end of a rule and a sample of the remedy on the other, they found that when a remedy was in balance with the patient's needs they got a pendulum reaction at the halfway mark on the rule. In other words the null point (where the polarities cancelled each other out) was at fifty percent on the scale. If the null point was closer to the patient's witness the remedy was too weak, and if the null point was closer to the remedy it was too potent.

In the 1930's doctors in Europe began to use this technique for diagnosis quite successfully. They would use a square or triangle as their balance rule, however, because these geometrical representations amplified the sensitivity and accuracy of the diagnosis when aligned with the earth's N-S magnetic field. First they found that a balance rule oriented north-south (magnetic) became "alive" with two resonant nodal points at each end, two other points about one seventh of the total length from each end, and one in the middle -- for a total of five points. They called this a One-Five Antenna.

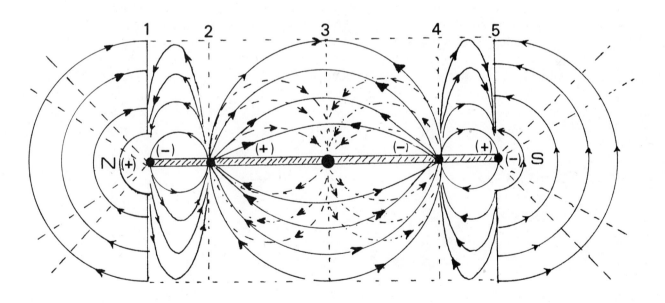

The Linear Antenna 1-5 pattern of a bar magnet showing field edges and resonant points between alternating (+) or (−) flows of magnetic lines of force as detected with a magnet-fitted rod or pendulum.

Rules had this kind of balance when two similars (similars occur when the witness and remedy resonate) were on opposite ends of a north-south aligned rule. Aligning any system north-south was found to automatically use the earth's magnetic field to amplify the magneto-dynamics of the system and thus its magnetic field. As they experimented further they discovered that a right triangle's base also became a One-Five antenna when the base was aligned to the east-west axis.

So these European radiesthesists began to use angles from the apex to measure the relationship between the patient's witness and the sample remedy. In this way they started to refer to conditions in terms of degrees. For instance, when a patient's witness balanced perfectly with a sample of perfect health (healthy plant) the angle of arc between the end of the rule and the null point would be forty-five degrees (45°).

Then someone else came along and tried orienting one side of the triangle north-south and found that the base, instead of being amplified by the earth's magnetic field, became excited by the earth's east-west flow of free atmospheric electrons. Whereas north-south alignment resulted in amplification of the magnetic field involving the balance, the east-west alignment amplified the radio-magnetic signals given off by the witness and sample as well as the magnetic field, a double amplification. It made diagnosis even more accurate as it was easier to find the null point of the balance.

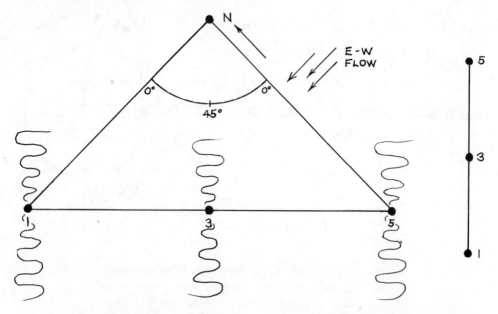

The excited antenna resulting from this new alignment was referred to as a One - Three - Five antenna because it had resonance nodes at both ends and at the null point in the middle.*

Then came the idea of using the diagonal of a square, which is made of two right triangles base to base. With the diagonal oriented east-west it became a linear One-Five excited antenna for amplifying the magnetic field of the patient's witness and the remedy. With the diagonal oriented north-south the whole square became excited and the diagonal became a One - Three - Five antenna. With a magnet-fitted pendulum a strong east-west flow could be detected sweeping across the face of the square. In fact, the east-west flow manifested even without a diagonal drawn in. A plain square with opposite corners aligned north-south would produce the east-west wavefront and double amplification associated with One - Three - Five antennas. However, with its sides oriented north-south it would act as a suppressor until a diagonal was drawn from corner to corner as with a pyramid.

Dr. Nebel of Lausanne, Switzerland, a medical radiesthesist famous for his accurate radiesthesic diagnoses of tuberculosis, is credited with developing the Square Balance as a diagnostic tool. Orienting the diagonal north-south he would place a sample of a sick patient, such as a bloodspot on a tissue paper, in the middle of the two intersecting diagonals and place various samples of the disease (in sealed vials) at the end point of the east-west diagonal until he found the disease sample which elicited a positive response through a pendulum. To find the properly balanced remedy he would place sample remedies at the east-west end point and test with a pendulum until he got a positive response. To amplify the field at the center of the square even further he would put a copper coil there. This excited the signal of the patient's witness. Dr. Nebel became very famous for diagnosing a patient without the patient even having to come to his office. The patient would phone up to say he wanted an appointment and the minute he walked in the door Nebel would tell him what his problem was!

Christopher Hills then came along with more improvements. He replaced the copper coil with a radioactive core (in homeopathic quantities) of heavy hydrogen (H3). Otherwise known as tritium, this core excited the wavefield of the witness in the middle even more than the copper coil. Hills added another improvement with the use of concentric squares one inside the other. This amplified the dynamics of the square form, thereby enhancing its diagnostic and treatment capabilities considerably. These concentric squares utilized another powerful force, that of the Great Pyramid form. Looking at the pyramid form from

---

* Specifics of resonance excited antenna and the east-west flow are dealt with in Vol. I and Vol. III of "The Supersensitive Life of Man Series."

above it looks exactly like a set of concentric squares within squares, one on top of the next. A two-dimensional symbolic representation of this form acts in the same way as the three-dimensional shape of a pyramid and holding one's hand over the square will energize the operator with positive green energy, just like the actual pyramid. All these features add up to the latest model of this balancing system, the Square Balance Equalizer.

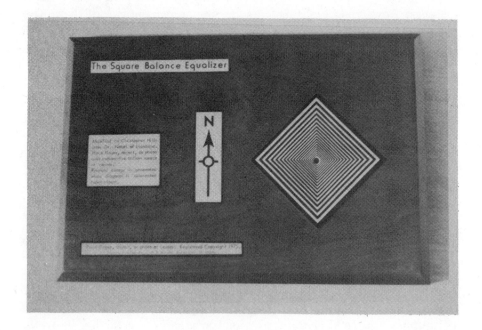

By incorporating pyramid dynamics into the design, Hills has turned the square balance into a radiation broadcaster as well as a diagnostic tool. With this device it is possible to determine what the problem is, what remedy is needed, and then broadcast the energetic pattern of the remedy to the patient. Some simple experiments that can be done with this device follow.

Radiating Remedies – By placing a picture or physical witness of a person (hair, fingernail, bloodspot) in the center of the square, a pendulum such as the Spectrum Mirror can be used to determine the nature of an imbalance as well as an appropriate remedy. As explained previously, when the operator's pendulum reacts positively to a particular witness/remedy combination via the Square Balance he then knows he has the right remedy. The next step is to repeat the procedure, this time to find out the proper amount and time of day for treatment.

The same procedure can be used with ourselves. Holding in hand a sample of an herb, for example black tea or marijuana, we can find out if it is in balance with our being by using a pendulum over the center of the Square. The principle at work is resonance radiation which is detected by the supersense.*

Balancing Our Nervous System -- With the corner oriented north-south, if we hold the fingers on one of our hands over the center of the Square for awhile there will be an automatic balancing of the sympathetic and parasympathetic nervous systems. Further, if we gently stroke the center of the Square a tingling sensation of energy passing onto the fingers is often noticeable. This energy is prana, otherwise known as psychic electricity or life force.

Daily planning – Many of us have a hard time achieving balance in our daily activities and often this is the result of doing the right thing at the wrong time (though doing the wrong thing at the wrong time is also frequent). Often if we only had our priorities straight we would avoid sticky problems. The Square Balance can be used to find out if a proposed activity is really in balance with our priorities. The pendulum will respond according to what the operator needs for staying in good psychological balance.

THE MAGNETRON

Those of us who think of ourselves as earthbound often forget that the ground we walk on is only one layer of the core of a multi-layered system of being. Surrounding this core are other layers of the earth's being such as the atmosphere, ionosphere and the magnetosphere. We are all polarized walking magnets constantly being affected unconsciously by the earth's magnetic flux which we can use for energizing ourselves. Through supersonic methods we can become aware of our bio-magnetic natures and learn how the earth's magnetic field can help us heal ourselves.

The magnetron is a good example. Developed in the early 1950's by a French radiesthesist named Servanx, the magnetron acts as a powerful radiesthesic link to the earth's magnetic field. Eight magnets of alternating polarity are arranged in a circle on its face and by aligning the device with the earth's north-south magnetic axis we "tune" it.

---

* For full description of the supersense and its ramifications see "Nuclear Evolution," by Christopher Hills, published 1977.

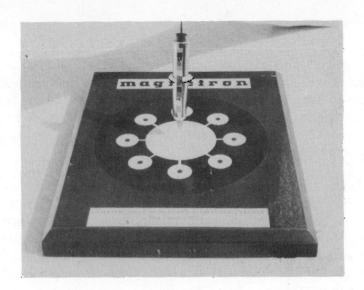

By placing a witness (i.e. a blood spot, hair, fingernail, etc.) of some person, plant or animal in the middle, we can create a supersensonic connection which will channel powerful healing energies into the patient, so powerful that we have to be careful of an overdose. The magnetron actually amplifies (with the power of the earth's magnetism) this wavefield of the witness in its center and will broadcast healing vibrations to anyone whose picture we place in the middle.

You can test this out for yourself by placing a friend's photograph in the middle and thinking "healing energy is going to that person." Use a pendulum to determine the length of the treatment. You can also use the magnetron to broadcast the vibrations of a remedy into the person's wavefield by placing a sample of a remedy on top of the photograph. The earth's field amplifies the healing vibrations and then focuses them on the person. To find the most suitable remedy just place a different choice over each of the eight magnets, hold a pendulum over the center of the magnetron and ask in your mind, "which remedy is needed?" The pendulum should begin to oscillate in the direction of the most suitable remedy. If it does not swing at all then you know none of the remedies are needed.

In any situation where there are several choices and you want to find out which is really best for your health, the magnetron is an excellent aid. The circle of magnets acts as a representation of the total spectrum of health. Place each of your different options in a different direction in this spectrum and hold your pendulum over a picture of yourself in the center. Ask your inner healer to point out the healthiest choice and soon the pendulum should swing in the proper direction. Remember that in the situation outlined above the answer you get is the best of those eight options. This does not mean that the choice is actually good for you. If the question you ask is "which of these do I need?" it might turn out that none of them are needed.

THE ULTIMATE INSTRUMENT

Behind the workings of all these wood and plastic instruments is the ultimate instrument, our consciousness. The secrets of the universe are hidden in its structure and once we turn on our nervous system and brain and awaken our supersense we will be able to tune in to the Creative Intelligence and discover what those secrets are. This work to amplify the signals of the nervous system with bits of wood, metal and plastic rods and pendulums is only a first step. Eventually we do not need the instruments and can create the healing circuits directly in our consciousness. We can do a number of experiments that prove our consciousness is the active ingredient.

If one of your plants is ailing take a photograph of it to use as a witness. Also get a photograph of the sun as a witness of vibrant health. Next hold the photographs together and vividly imagine the healing energy of the sun streaming into the plant. Repeat this exercise five minutes a day for a week and soon you should notice a change in the plant. Use another ailing plant as a control and compare the difference.

Supersensonic instruments are tuned to forces which are used as carrier waves for irradiating someone or something with healing vibrational patterns. Our conscious attention and sense of conviction act as the catalyst for triggering these cosmic forces as carriers. We are learning that all the energy systems in our vehicle are hooked up to a universal field of primary light which is crystallized by our chakras and cerebro-spinal system into our cellular structures. We are also learning that each of us is our own best expert in awakening our super-sensitive circuitry and expanding our awareness to include the hologram of the larger whole. The ancient diviners who awakened this faculty roamed the universe and were able to discover the secrets of life in their consciousness. Einstein travelled around the universe on a beam of light in his consciousness. Even though this quality of insight is rarely seen in the majority of the human race, we all have the potential to open up these channels of direct communication with the cosmos. The ancient science of direct perception is now coming of age. Through the new science of supersensonics we begin to see for the first time how really vast is the power of our consciousness -- our Inner Healer.

REFERENCES FOR SECTION THREE

BOOKS

1. Archdale, F., Elementary Radiesthesia, British Society of Dowsers, London, 1966.
2. Beasley, Victor, Dimensions of Electro-Vibratory Phenomena, University of the Trees Press, Boulder Creek, 1974.
3. Besteman, Theodore, Crystal Gazing, Rider & Son: 1924.
4. Bieler, Dr. Henry, Food is Your Best Medicine, Random House: New York, 1968.
5. Blair, Lawrence, Rhythms of Vision, Schocken Books: New York, 1976.
6. British Society of Dowsers, Radiesthesic Approach to Health and Homeopathy, London, 1961.
7. De France, Henry, The Elements of Dowsing, G. Bell: London, 1959.
8. Dietrich, C., M.D., Pendulum Diagnosing, Bruce Copen: Sussex.
9. Friedell, Aaron, M.D., May You Live in Health, Friedell: Minneapolis, 1968.
10. Lama Anagarika Govinda, Foundations of Tibetan Mysticism. Rider: London, 1960.
11. Hills, Christopher, Nuclear Evolution, Centre Community Publications: London, 1968.
12. Hills, Christopher, Christ-Yoga of Peace, Centre Community Publications: London, 1970.
13. Hills, Christopher, Conduct Your Own Awareness Sessions, New American Library: New York, 1970.
14. Hills, Christopher, Supersensonics, University of the Trees Press: Boulder Creek, 1975.
15. Hills, Christopher, Rays From the Capstone, University of the Trees Press: Boulder Creek, 1976.
16. Hills, Christopher, Instruments of Knowing, University of the Trees Press: Boulder Creek, 1975.
17. Hoyle, Fred, Ten Faces of the Universe, W.H. Freeman & Company: San Francisco, 1977.
18. Hyslop, James, Enigmas of Psychical Research, Turner & Company: Boston, 1906.
19. Jenny, Hans, Cymatics, Basilius Press: Basel, 1967.
20. Jung, Carl G. (Ed.), Man and His Symbols, Dell Publishing Company: New York, 1964.
21. Macbeth, Noel, Radiational Physics Notes, A-A-P Essex (Sec. 1-4).
22. Maltz, Maxwell M.D., Psycho-Cybernetics, Pocket Books: New York, 1960.
23. Maslow, Abraham H., The Psychology of Science, Henry Regnery Company: Chicago, 1966.
24. Maslow, Abraham H., Religions, Values and Peak Experiences, Ohio State University Press: Columbus, 1964.
25. Maslow, Abraham H., Toward a Psychology of Being, Van Nostrand Reinhold: New York, 1968.
26. Massy, Robert, Alive to the Universe, University of the Trees Press: Boulder Creek, 1967.
27. Mermet, Abbe, Principles and Practice of Radiesthesia, Vincent Stuart, 1935.
28. Mishra, Rammurti M.D., The Fundamentals of Yoga, The Julian Press: New York, 1959.
29. O'Byrne, F.D., Reichenbach's Letters on Od and Magnetism (1852) Health Research: Mokelumne Hill, 1964.
30. Ostrander and Schroeder, Psychic Discoveries Behind the Iron Curtain, Prentice Hall: New Jersey, 1970.
31. Puharich, Andrija, Beyond Telepathy, Anchor Books: New York, 1973.
32. Piaget, Jean, Biology and Knowledge, University of Chicago Press, Chicago, 1971.
33. Richards, Guyon, M.D., The Chain of Life, Health Science Press, 1954.
34. Rozman, Deborah, Meditating With Children, University of the Trees Press: Boulder Creek, 1975.
35. Rutherford, Adam, Pyramidology Books 1-3, Institute of Pyramidology: London.
36. Strutt, Malcolm, Theory and Practice of Using the Pendulum, Centre Community Publications: London, 1971.
37. Samuels, Mike, M.D. and Nancy, Seeing With the Mind's Eye, Random House/Bookworks, New York/ Berkeley, 1975.
38. Schwenk, Theodore, Sensitive Chaos, Rudolf Steiner Press: London, 1965.
39. Thomas, Lewis, The Lives of a Cell, Bantam Books: New York, 1974.
40. Trinder, W.H., Dowsing, British Society of Dowsers: England, 1939.
41. Tompkins, Peter, Secrets of the Great Pyramid, Harper and Row: New York, 1971.
42. Tompkins, Peter and Bird, Christopher, The Secret Life of Plants, Avon: New York, 1973.

43 Wethered, Vernon D., Radiesthesic Approach to Health and Homeopathy, British Society of Dowsers, 1961.
44 Wethered, Vernon D., The Practise of Medical Radiesthesia, L.N. Fowler and Company: London, 1967.
45 Wentz, W.Y. Evans, The Tibetan Book of the Dead, Oxford University Press: London, 1927.
46 Wentz, W.Y. Evans, Tibetan Yoga and Secret Doctrines, Oxford University Press: London, 1935.
47 Wentz, W.Y. Evans, Tibet's Great Yogi Milarepa, Oxford University Press: London, 1928.
48 Wilhelm, Richard, (trans.) The Secret of the Golden Flower, Routledge and Kegan Paul: London, 1931.
49 Worrall, Olga and Ambrose, Explore Your Psychic World, Harper and Row: New York, 1970.

ARTICLES

Douglas, Nick, "The Basis of Buddhist Vajrayama Tantric Meditation Practice," Chakra Vol. 2, Kumar Gallery, New Delhi, 1971.

Douglas, Nick, "Eight Mudras and Mantras From the Mahakarunacitta-dharani" Chakra Vol. 3, Kumar Gallery, New Delhi.

Douglas, Nick, "Ajit Mookerjie," Chakra Vol. I, Kumar Gallery, New Delhi, Winter, 1970.

Chen Yogi, C.M., "Annuttarayoga, the Wisdom Drops and Vajra Love," Chakra, Vol. I.

TAPES

Hills, Christopher, "Water," Centre Community Publications.

Hills, Christopher, "Temple of Awareness," Centre Community Publications, London.

Hills, Christopher, "Rumf Roomph Yoga No. 9 -- the Great Yogic Breath," University of the Trees Press, 1975.

Hills, Christopher, "Rumf Roomph Yoga No. 16 -- Creative Conflict," University of the Trees Press, 1975.

Hills, Christopher, "Supersensonics Seminar -- Tapes Numbers 1-6," University of the Trees Press, 1976.

Hills, Christopher, "Vision -- Open Session No. 96," University of the Trees Press, 1976.

Hills, Christopher, "Yoga in Genesis," Centre Community Publications, London, 1971.

Hills, Christopher, "Fourth Lecture on Meditation," Centre Community Publications, 1971.

Hills, Christopher, "Organ of Perception," Centre Community Publications, London.

Hills, Christopher and Dr. Woidich, "Endocrine Glands," Centre Community Publications, London.

EPILOGUE

TOWARD A SCIENCE OF CONSCIOUSNESS

FUTURES RESEARCH

Speculations about conceivable alternative futures for the family of humanity, deduced from in-depth analyses and projections of present worldwide trends, have resulted in alarming diagnoses with dire, increasingly deterministic prognostications if these trends are not slowed (for example, population growth), halted (for example, pollution), or reversed (for example, nuclear weapons proliferation). Some rather big changes need to be made at the individual, community and planetary levels. The future, however, lies in the thoughts, words and deeds of each of us. With the range of choices available multiplied over four billion times and further complicated by environmental unknowns, the problem of accurately forecasting (prophesying) the possible alternative futures for humanity as a whole becomes complex indeed. Yet our volition is constrained by the environmental circumstances in which we find ourselves, circumstances which now increasingly limit our available choices. Consequently, the importance of deliberate choice and its origin, human consciousness, have become critical. Like a nuclear pile, we have reached the stage of "critical mass" and are ready for nuclear transformations in our consciousness.

Several prominent writers engaging in "futurology" emphasize that the planetary problems facing us cannot be overcome without revolutionary changes in humanity itself. Dr. Robert Heilbroner, in *An Inquiry Into the Human Prospect,* although despairing that "there seems no hope for rapid changes in the human character traits that would have to be modified to bring about a peaceful, organized reorientation of life styles," nevertheless admits the possibility that "a post-industrial society would also turn in the direction of inner states of experience rather than the outer world of fact and material accomplishment." While he assumes that such exploration would result in a return to tradition and ritual (negative blue level) and despite his despair of the inability to change human nature this, paradoxically, is where he places his hope. Using the metaphor of Atlas, he calls for the fortitude and will of the inner person to come forth and save the world (positive blue level).

In *Between Two Ages* Dr. Zbigniew Brezinski, National Security Adviser to President Carter, speaks of an emerging global consciousness arising "as a natural extension of the long process of widening man's personal horizons." He asserts that "the sense of proximity, the immediacy of suffering, the globally destructive character of modern weapons all help to stimulate an outlook that views mankind as a community." He notes that the frameworks of established cultures, traditional religions and national identities is disintegrating with new frameworks yet to take coherent form. However, as we have seen, a new framework *is* taking form. A world culture practicing a spiritual science in which everyone has rights as a world citizen is evolving.

In *World Without Borders* Dr. Lester Brown calls for a new humanist social ethic (a set of principles enabling the world society to function and survive as a huge, self-governing organism). Noting that such an ethic would have far reaching implications for the behavior of individuals and communities alike, he contends that "its adoption would certainly result in new life styles and a new society, one far different from the one we now know." All three of these future thinkers, then, place their hopes for our future in terms of human consciousness, whether it be the selfless spirit of Atlas, a global consciousness or a new ethic of humanism.

There is a growing body of evidence to support the view that just such a hoped for transforming process is occurring. There is a new spiritual scientific movement underway which may have far greater impact on the futures of all of us than all man-made constructs put together, for it is ultimately the source of them. This movement is toward a unified spiritual science of consciousness.

THE SCIENCES ARE MERGING

The whole edifice of the classical world view is crumbling like the walls of Jericho under the clarion call -- WHAT IS CONSCIOUSNESS? The question of consciousness is a Sphinx to modern materialistic science, for it is inscrutable and impenetrable. We can speak forever about consciousness but we can never say what consciousness is. As more of us recognize the fundamental importance of consciousness, many of us realize that to study it we have to take a deep look into ourself. That is the only way to really know what consciousness is. In doing this we fulfill the Biblical injunction: MAN KNOW THYSELF. No longer onlookers, we become participators in purifying ourself of separation and expanding our consciousness to realize spiritual union, the inherent oneness of us all.

By participating as well as observing we become psychoscientists in experiential experimentation combining reason with intuition through techniques of direct perception. As we try to answer the Sphinx-like question of consciousness and other questions which inevitably arise from it, the sciences are undergoing radical transformations down to their very philosophical roots. As a result, many new sciences -- the psychosciences -- are exploring consciousness and are uniting into an interdisciplinary study due to a profound change in scientific thinking at the precise moment it is most needed – NOW! Just as Copernicus and Einstein changed traditional, orthodox scientific views about the universe, Darwin and De Chardin changed our view of our place in it, and Freud and Maslow changed our view of ourself, so may scientific investigation and technical applications to expand consciousness radically change individual awareness and world thought. The blending of inner with outer, real with ideal, micro with macro, science with spirituality and self with Self will become more closely achieved as we turn our awareness, concentrated attention and highly developed tools of investigation to the study of consciousness, beginning with ourself.

Our great planetary being, humanity, appears to be in the throes of a major collective catharsis, a metamorphosis, a rebirth. The Earth is convulsing with labor pains. Around us ring cries that we are entering a New Age, where anything is possible. The scientific study of consciousness promises to bring that New Age by helping us to purify and increase our awareness, spinning us ever higher together on the grand evolutionary spiral of expanding consciousness.

THE PSYCHOSCIENCES

As we have seen, consciousness unites mind with "matter" and is the center of the new psychophysics of Wheeler, Young, Sarfatti, Wolf and Hills. Consciousness underlies the other new psychophysical sciences of psychoptics, psychophysiology, psychobiology, psychoneurology, psychochemistry, psychopharmacology, psychotronics and parapsychology. Consciousness lies in the heart of the new psychosocial sciences of psychosociology, psychopolitics and psychohistory.* None of these new sciences existed more than a decade ago. They have all arisen from the study of human consciousness by human consciousness, the foremost study of them all -- the study of Self.

---

* Psychohistory is the study of the past from the standpoint of human expression based upon the visionary experience and the evolution of consciousness. In an excellent book on the subject, "The Transformative Vision," José Argüelles notes that:

> History is a dream -- the dream of reason. As long as man believes in this dream and seeks to acquire an historical identity, he remains unconscious of the fact that he is a bridge between the cosmic realms of heaven and earth. His only escape from this fatal circle is to wake up from this dream and realize a cosmic, mythic, and fundamentally timeless identity.

With the tools and principles of supersensonics as a guide, the psychoscientists in these new disciplines can add the power of intuition to complement and amplify their reasoning. Years of trial and error research can be eliminated by tuning in to the universal intelligence with the instruments of knowing designed and invented by Christopher Hills and asking simple, direct questions with yes/no answers. The results of these divining experiments in direct perception can then be verified with conventional laboratory apparatus and techniques. Because of the tremendous breakthroughs possible from the study and practice of supersensonics we may soon see other new psychosciences such as psychogeology. This involves the locating and assaying of elements, minerals etc. through radiesthesic methods, as well as finding and "dating" archeological discoveries. Psychomedicine combines psychosomatic medicine with radionic techniques of analysis, diagnosis and treatment.

In short, all the many specialized sciences we know today will be united and totally transformed by supersensonics. Supersensonics is fundamentally grounded in consciousness, and consciousness is the divine spiritual energy which unites mind with "matter," mental and spiritual vibration with physical form. This heals the mind-body dichotomy that has long divided science. Through supersensonic techniques we can directly contact the universal field which unites everything. The spiritual science of supersensonics will unite all the sciences like rays of healing light radiating from a common, central, spiritual sun – *PURE CONSCIOUSNESS*.

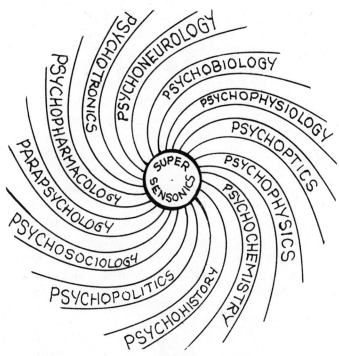

CONSCIOUSNESS RESEARCH

At the University of the Trees,* a consciousness research center nestled among the towering redwoods of Boulder Creek, California, a unique experiment in community living has been underway for several years. The primary curriculum is the three-year course "Into Meditation Now" (being taken by correspondence by people all over the world), though a host of programs and courses as diverse as the intellectual backgrounds of the student/faculty who teach them is also available by correspondence. The University of the Trees is one of the first degree granting institutions without walls offering bachelor's, master's and doctoral degrees in consciousness research and related fields. Degree programs are tailored to the needs and aspirations of each individual with credit given for substantiated work experience related to the degree. Some of the suggested majors in the graduate division include:

Psychology of Consciousness	Social Group Dynamics
Physics of Consciousness	Consciousness and Healing
Philosophy of Consciousness	Communities and Self-Government
Poetry of Consciousness	New Foods for Human Survival
Alternative Energy Systems	Awareness Enhancement in Education
Psychophysiology	Cosmic Mathematics
Psychotronics -- Supersensorics	Colors and the Mind

The University of the Trees is also a residence community offering courses and intensified training in consciousness research. In order to qualify for residence, prospective new members have to be taking the three-year meditation course and be interviewed and accepted by the entire community family. This is an experiment in group consciousness where selfless service is a way of life. All of us are self-supporting and volunteer from our own hearts considerable time and energy to community endeavors, such as producing the instruments of knowing and publishing books which are the result of the student/faculty research through the university press. We emphasize quality and depth in our work on ourselves and in our research, and accept those who are open to growth and change. Profits from the businesses are being accumulated and set aside for the eventual establishment of the World Peace Center described in Chapter Six. The commitment here is to total openness, to discovering ourselves in the mirrors of each other, and to everyone's enlightenment, gradually building toward a truly united group spirit which can expand to include all beings everywhere. We are just ordinary people from all walks of life committed to doing extraordinary things together for the good of the whole, that ONE who unites us all.

*The trees refer to the Biblical tree of knowledge (representing the nervous system) and the tree of life (representing the nerve dendrites of the brain).

PART OF THE GROWING FAMILY AT THE UNIVERSITY OF THE TREES
REAR (left to right): *Christopher Hills, Norah Hills, Ann Ray, Dave Edwards, Jeff Goelitz, Gary Buyle, Roger Smith, Phil Allen, Ted Saunders, Michael Hammer.* MIDDLE ROW (standing): *Dan Hime, Sue Welker, Sue Belanger, Wendy McFadzen, Diane Vandewall.* FRONT (seated): *Pam Osborn, Robert Massy, Rod Glasgow, Richard Welker, Stephanie Herzog, Debbie Rozman.*

INDEX

Acupoints: 14.
Acupuncture: 18, 53, 103, 117.
Adamenko, Victor G.: 15.
Adrenaline: 25, 32, 108.
Aesculapius: 64.
Argüelles, José & Miriam: 20.
Arp, Halton: 124.
Assagioli, Roberto: 5, 147.
Atom: 22, 57, 60, 89.
Aura: 17, 19, 20, 30, 43, 45-48, 74, 236.

Baudoin, Charles: 149.
Beasley, Victor: 7, 8, 57, 92, 108.
Becker, Robert O.: 10, 14.
Belizal: 195.
Bell, Alexander G.: 102.
Biocommunication: 9.
Bioenergy: 2, 11, 12.
Biofeedback: iii, 2, 232, 234.
Biogravity: 126.
Bioluminescence: 8, 9, 56, 152.
Bioplasma: 56, 109, 118.
Black Hole: 121-126.
Bovis, Andre: vi, 201.
Branley, Dr.: v.
Bradley, Matthews O.: 28.
Breathing: 52, 57, 62, 73, 75-81, 108, 132-135, 177-180.
Brodsky, Greg: 60.
Buddhism: 4, 23, 27, 87.

Caduceus of Mercury: 63-66.
Cancer: 14-15, 28-9, 202, 205-206.
Capra, Fritjof: 6, 87.
Cayce, Edgar: 20.
Cells: 7-9, 15, 17, 27, 29, 58, 70, 120, 138-139, 178, 221.
Center: 20-23, 70, 74, 79, 126-128, 130, 138, 152, 160.
  -consciousness: 74.
  -constitution: 161, 162.
Centre Community: ii.
Chakra: 22-27, 32, 33, 47-50, 62-66, 68, 129, 213, 237.
  -dynamics: 169.
Chanting: 134, 168, 173, 175-177, 181-184.
  -group: 182.
  -troupad: 183.

Chaumery: 195.
Chladni: 172, -plate: 185.
Christianity: 4, 23, 87.
Clairvoyance: 20, 23, 220, 226-228
Coffer: 114, 198, 199.
Color: 24, 30, 31, 42, 133, 134, 218, 227, 228, 236, 245, 258,
  -healing: 257, 259.
  -physiology of: 25-30.
  -psychology of: 30-42.
Communication: 102, 152-155, 183.
Concentration: 2, 26, 50, 79, 233.
Conflict: 160, 167, 211.
  -creative: 152-156, 210.
Cosmic energies: 182.
Crystal oscillators: 186.

D'Arsonval, Dr.: v.
Darwin, Charles: 139.
Davis, Albert: 60.
Davis, Roy A.: 10.
De Chardin, Pierre Teilhard: 5, 84, 139.
De La Warr, George: 15, 58, 260.
Delgoff, Eugene: 116.
Descartes, Rene: 85, 88.
Deprogramming: 182.
Detachment: 175.
Disease: 14, 45, 169, 249, 252-254, 265, 271.
Divining: iii, iv, 50, 169, 274.
DNA: 7, 17, 25, 58, 61, 103, 120, 209.
Dorje: 220, 226.
Douglas: John: 142.
Dowsing: 169.
Drbal, Karl: 203.
Dubrov, Dr. Alexander: 8.
Dumitrescu, Ioan: 14.

Ego: 36, 138, 140, 152.
Einstein, Albert: v, 2, 55, 85-91, 123, 139.
Electromagnetic: 9-11, 21, 55, 64, 88, 168.
Electrophotography (Kirlian): 3, 13, 14, 17, 46, 118, 151.
Endocrine System: 17, 27, 31.
  -adrenal: 25, -parathyroid: 26.,

-pineal: 27, -pituitary: 27, 32, -thyroid: 26.
Energy: 2, 6, 7, 14, 18, 22, 23, 25, 43, 44, 47, 66, 70, 131, 139, 153, 159, 160, 168, 170, 191, 195-199, 205, 206, 249.
Energy Broadcasting: 169.
Ether: 20, 21, 53, 55.
Eye: 21, 26, 27, 50, 79, 95, 98, 112, 127, 131, 166, 168, 199, 200, 207, 216-218.
  -beam: 112, 131.
  -Ideal Eye: 85.

Fiberoptics: 142, 144.
Form: 109, 166-170, 184-191, 264.
Freud, Sigmund: 140.

Gaiken, Mikhail: 14.
Garrett, Eileen: vi, 109.
Geller, Uri: 106.
Gravity: 187-188.
Great Pyramid: iv, 113, 192-193, 195-197, 199, 201-202, 231, 278.
Great Seal of the U.S.: 199, 200.
Great Yogic Breath: 177, 181.
Group Consciousness: ii, 150-151.

Hameroff, Stuart Roy: 16, 118.
Hampden-Turner, Charles: 149.
Han, Kim Bong: 16, 56, 116.
Health: 10, 11, 14, 27, 45, 178-180, 205, 206.
Heisenberg, Werner: 90.
Hills, Christopher: 21, 26, 42, 48, 52, 58, 70, 71, 86, 91, 91, 95, 100, 128, 130, 140.
Hills' Positive Green Pendulum: 205, 206.
Hinduism: 4, 23, 87.
Hologram: 56, 114-115, 119-121, 145, 231, 247.
Holographic Chip: 182.
Holy Bible: 64, 65, 152.
Hurkos, Peter: 109.

I Ching: iv, 231.
Ideal I: 85-86, 90, 94.

Images: 31, 35, 40-41, 96-97, 114, 121, 125, 132-135, 137, 144, 146, 148-150, 166, 168, 171, 175, 208-209, 215, 226 (Self) 119, 149, 209-210, 211, 212, 220, 226-117, 230-231, 247.
Imagination: 2, 21, 40, 41, 42, 128, 132, 141, 145-146, 167, 170, 171, 207, 213, 214, 220, 224, 230.
 -creative: 168.
Interference patterns: 171, 173.
Intuition: 170.
Inyushin, Victor: 14, 56, 112, 118-119.
Ionization: 56, 57, 60, 108, 109, 178.
Inner Listener: 170.

Jeans, James: 87.
Jung, Carl: 140.

Kamensky, Uri: 102.
Kaufmann, William J.: 88, 121,
Kilner, Walter: 45.
Kim, Young S.:
Kirlian, Semyon & Valentina: 11.
Kogan, I.M.: 102.
Krippner, Stanley: 10.
Krmssky, Julius: 111.
Kundalini: 70-72, 262.
Kundt, August: 189.

Land, Edwin: 30.
Larvaron, M. iv.
LePoiree, Dr.: iv
Life Energy: 3, 53, 54, 72, 75-76, 151, 154, 160, 195.
Light: 2, 3, 5, 17, 23, 17-30, 42, 55, 60, 74, 88, 116, 118, 170, 179, 200, 209, 245, 258.
Liquid Crystals: 8, 58, 61, 143, 178

Macbeth, Noel: v.
Mager, Henri: iv.
Magnetism: v, 8, 242, 243.
Magic: 42, 212.
Mann, Felix: 53.
Mantra: 168, 173-176, 224.
Maslow, Abraham: 5.
Masons: 199.

Massy, Robert: 3, 21, 59, 59.
Maury, Dr. E.A.: iv.
McCready, Dr.: v.
Meditation: 2, 12-23, 43, 71, 80, 107, 108, 150, 196.
Merton, Thomas: 5.
Micro Algae: 157-159.
Mind's Eye: 168.
Mirroring: 153.
Mishlove, Jeffry: 14, 17, 104, 118, 125.
Mishra, Rammurti S.: 103, 110.
Mitochondria: 58, 59, 118.
Mitosis: 8, 116, 120.
Moreno: 154.
Moss, Thelma: 12, 15, 102, 111.
Muses, Charles: 90, 121, 128, 150.
Mysticism: 4, 20, 270.

Negative Green: 196-197, 201-202, 204-206.
Newton, Isaac: 139.
Nikolaev, Karl: 102.
Nirvana: 4, 5.
Nodal Points: 15, 58, 125, 190.
Nuclear Evolution: 5, 33, 42, 56, 58-59, 73-74, 79, 98, 100, 118, 120, 140, 150-151, 157.
Nucleus: 20, 21, 57. 138, 150-152.

Om: 20, 21, 176, 177, 179, 180, 183
Ostrander, Sheila: 11, 14, 56, 112, 132.
Ott, John N.: 27-29.
Oyle, Irving: 16.

Paraphysics, Radiational: 91.
Parapsychology: i, 86, 103, 286.
Pavlita, Robert: 111.
Pendulum: 48-49, 135-136, 205-206
Perception: 31, 36, 40, 85, 97, 148, 153, 162, 175, 230.
Pettman, Ralph: 155, 157.
Physics: vi, 6.
 -philosophy of: 87-90.
 -subatomic: 89.
Pi (Phi): 193, 194.
Pi-Ray: 113, 114, 168, 197-199.
Picasso, Pablo: vi.
Planetary Being: 138-139, 150, 154
Planetary Consciousness: 139, 148.

Plants, vibrational effect on: 204.
Plato: 194.
Plexus: -cardiac nerve: 26.
 -sacral nerve: 25.
 -solar nerve: 25.
Positive Green: 51, 168, 195, 197, 204-206, 273.
Prana: 52-53, 56, 67, 72, 75, 131, 195, 205-206, 279.
Presman, A.S.: 9, 10.
Pribram, Karl H. 115.
Projection: 153.
PSI Communication: 86, 102-105, 115.
Psychic Electricity: 22, 23, 63, 65, 67, 197, 179.
Psychobiology: 139, 286.
Psychochemistry: 31.
Psychodrama: 154.
Psychoendocrinology: 31.
Psychohistory: 286.
Psychomedicine: 31.
Psychoneurology: 31.
Psychophysics: 86, 126, 286.
Psychoptics: 95, 97, 286.
Psychophysiology: 95, 286.
Psychopolitics: 155, 286.
Psychosciences: 31, 286-287.
Psychosociology: 149, 286.
Psychosphere: 141.
Psychotronics: 86, 102, 110, 112, 113, 131, 261, 286.
Puharich, Andrija Dr.: 109, 110, 112
Pure Consciousness: 5, 42, 70, 81, 124, 162, 175, 207, 225, 287.
Putoff, Harold: 105.
Pye, Lucian: 155.
Pyramid Energy: vi, 195, 201.

Qi (Chi): 15, 53, 54, 56, 60, 118, 131, 195, 205.
Quakerism: 5.

Radiation: 2, 9, 20.
Radiational paraphysics: vi, 91.
Radiesthesia: iii, iv, vi, 91, 169, 231-233, 234-235, 239, 246, 248, 250.
Radiomagnetic: 169.
Radionics: 91, 169, 248-149, 251, 254, 259, 260, 272.
Rays of Life: 168.

Regeneration, Physical: 10.
Regnault, Dr.: iv.
Reiser, Oliver: 7.
Rendel, Peter: 67.
Resonance: 176, 184, 234, 241.
Rhine, J.B. & Louisa: 105.
Richards, Dr. Guyon: v.
Ronchi, Vasco: 95-100.
Rozman, Deborah: 149.

Sarfatti, Jack: 121, 125, 126-127.
Schroeder, Lynn: 11, 14, 56, 112, 132.
Scrying: 227-229.
Seeing: 48, 84, 85, 95, 149, 189, 216, 219.
Self-image: 182.
Self-organization: 10, 139, 141.
Sharkey, Nancy O.: 28.
Siegel, Eric: 144.
Sienko, Michell J.: 60.
Singularity: 122, 126.
South, John: 29.
Sphinx: 2, 167, 285.
Spleen Center: 25.
Studitsky, Alexander: 11.
Sufism: 5, 23.
Sun: 3, 7, 21, 28, 55, 56, 60, 66, 68, 69, 78, 79, 140, 158, 199.
Superseeing: 137.
Supersensing: 168, 169, 230-237.
Supersonics: i, 20, 63, 70, 91-92, 95, 97, 128, 130, 137, 141, 170, 198, 287.
Symbols: 41, 79, 147, 176, 223, 243, 244, 264.

Tao: 4, 54.
Thought Patterns: 172.
Thought Waves: 2, 10, 49, 86, 132, 240.
Toben, Bob: 126.
Torus: 128, 130
Turenne, Louis: iv, vi, 231, 241, 245, 246, 263.

Underhill, Evelyn: 4.
Unified Field: 167, 169
Universe: 2, 6, 15, 33, 85, 123, 139, 170.
Universal Field: v, 138, 141.

University of the Trees: 24, 144, 154, 183, 199, 205-206
Ultraviolet: 27, 28, 30.

Verba, Sydney: 155.
Vibration: v, 2, 21, 23, 62, 73, 92, 102, 126, 166-171, 175, 177, 184, 188-191, 197, 204, 220, 237.
 -disease causing: 169
 -fundamental: 183
 -patterns of: 166-170, 172-173, 184-191.
Vire, Dr.: iv
Vortex: 22, 68, 122, 128, 131.

Wexler, Jesse S.: 16
Wheeler, John A.: 90, 125.
White, John: 5, 148.
Wigner, Eugene: 90
Wilson, D.H.: 10
Witness: 199, 280
Wolf, Fred: 125
World Culture: 137-138
World Peace Center: 161-162, 288.
Worrall, Olga: 105.
Worsley, J.R.: 13.
Wright, Mrs. Dudley: v

Yoga: 22, 23, 24, 48, 52, 63, 65, 70, 72, 78, 79, 104, 110, 127, 175, 197.
Young, Arthur: 128.
Youngblood, Gene: 141, 143, 145, 156.

BOOKS   ORDER FORM

UNIVERSITY OF THE TREES PRESS

Publishers of practical spiritual guides, scientific books and Correspondence courses

SUPERSENSONICS, by Christopher Hills   15.00
The Diviner's bible and encyclopedia that describes actual methods of measuring psychic electricity (prana) and the ways ancient masters and civilizations arrived at advanced knowledge of perception which shows you how to communicate with plants, crystals, atoms, Cosmic intelligences or your true Self. A mind-bending book.

ALIVE TO THE UNIVERSE!, by Robert Massy   7.50
A physicist explains Supersensonics in simple layman's language, and gives step-by-step instructions on how to divine for lost objects, people, water, minerals, health, etc. Your vast potential for multi-dimensional awareness awaits unfolding through this book. An illustration a page.

NUCLEAR EVOLUTION: THE DISCOVERY OF THE RAINBOW BODY
by Christopher Hills   Full color cover paperback - $9.95
This book is perhaps one of the most controversial and authoritative documents on spiritual, scientific and political man every published. Researching the role of Consciousness in the human species, Nuclear Evolution provides the unique synthesis of scientific evidence and spiritual guidance. It explores light synthesis by man, planetary political consciousness, aura colors, chakras, Einsteinian theory, Jungian typologies, the I Ching, time, space and human drives. Imitation leather, embossed Library Edition - $18.95.

RAYS FROM THE CAPSTONE, by Christopher Hills   4.95
After 20 years of in depth research into pyramid energies, we now have a book which contradicts many fantasies about meditation and pyramid power, shows the positive and negative uses of pyramid energies and explains the Pi-ray orgone accumulator coffer, a "natural" power dynamo invented by the author which allows you to use pyramid energy safely for growing healthier plants, changing emotional states, zapping yourself with energy, and purifying consciousness.

SUPERSENSONIC INSTRUMENTS OF KNOWING, by Christopher Hills   1.95
This book provides instructions on the use of various Supersensonics tools, rods, pendulums, etc, in a complete catalog of biofeedback instruments and their uses.

HILLS' THEORY OF CONSCIOUSNESS, by Robert Massy   5.95
A student of a master of consciousness describes his own development in a group of 15 selected students of Nuclear Evolution. This book contains simplified accounts of inspired research of the actual structure of Consciousness and its many drives which physicist Massy believes will dominate the next 500-1000 years of man's history.

WHOLISTIC HEALTH AND LIVING YOGA, by Malcolm Strutt   7.95
A new wholistic approach to the spiritual science of total health which includes the application of yogic principles to daily living situations, complete self-examinations and encounters, yoga postures, meditations, breathing and energizing methods, and other practical techniques for spiritual stimulation.

JOURNEY INTO LIGHT, by Ann Ray   7.95
By weaving a woman's story of spiritual struggle and unfolding with wondrous insight of time, space and consciousness a delightful and informative book has been created. Seemingly remote modern scientific breakthroughs are related to everyday life and problems. Different from the usual books on spiritual teachings, this book on the principles of Nuclear Evolution is not abstract and untested, but is written out of real experience and proven in the laboratory of the heart.

UNIVERSAL GOVERNMENT BY NATURE'S LAWS, by Christopher Hills   7.95
A radical breakthrough theory of government to answer the pressing problems of today, this book shows how to create a new social and political order through organised self-government. Commended by Bertrand Russell who wrote the introduction to the first edition (published in limited quantity) and said that these methods provide the only possible way of combining discipline with freedom and freedom with organization, we now have a viable approach to a new constitution based on nature's models.

INTO MEDITATION NOW: A COURSE OF STUDY, by Christopher Hills   45.00
This cost covers the registration and introduction to the comprehensive three-year course of study that enables you to make the philosophy of Nuclear Evolution a reality in your direct experience. Write for more details.

MEDITATING WITH CHILDREN, by Deborah Rozman   5.95
The first of its kind! A delightful teaching book that brings the great art and science of meditation and conscious evolution to children of all ages, this workbook is being used in classrooms throughout the country as a non-religious text in centering and awareness development.

These Prices Subject to Change Without Notice

SUPERSENSONICS TOOLS AND INSTRUMENTS

The following divining tools mentioned in the text are aids for contacting intelligences or energies, and are not substitutes for spiritual work on ourselves.

AURA PENDULUM: A low cost pendulum for determining a person's aura and the psychic atmosphere that surrounds them. $6.00. An excellent accompaniment is "The Rainbow Aura Pendulum Booklet" for $1.00.
THE OSIRIS: A glass pendulum containing mercury which is used to read people's thoughts, divine their I.Q., and tune into intuitions. A beauty at $16.00.
THE MERKHET: Has the same use as the Osiris but the mercury is contained in an unbreakable body for use in a tougher environment. $12.50.
ROOMPH COIL: For those who wish to raise their Kundalini energy. $83.00.
HILLS POSITIVE GREEN PENDULUM: Is specifically tuned to life-force or prana and is therefore used for checking people's health and vitality, the nutritional value of foods and herbs, and checking pyramid energies. $15.00.
PI-RAY ORGONE ENERGY ACCUMULATOR COFFER: "It materializes your imaginings," said one user. "It has helped bring about some remarkable events, including healing, selling properties, finding houses and cars for people at the right price, and increasing one's spirituality." $79.00.
EGYPTIAN ANKH DIVINING ROD: Is a type of divining rod that made the ancient Egyptians an advanced civilization. For use in the field. $8.00.
TURENNE MAGNET-FITTED ROD: A very selective divining rod, used for outdoor work in much the same way as Moses and Jacob used their rods. Excellent for water, mineral and lost objects locating. $22.00.
TURENNE MAGNET-FITTED PENDULUM: Called the Rolls Royce of pendulums because it has a wide range of selectivity and is specially suited for use with the Turenne Rule for detecting atomic and chemical substances. The researcher's pendulum. $20.00.
TURENNE FUNDAMENTAL RAY DISC: For chemists this disc takes the place of expensive analytical instruments. For those concerned about health and self-healing, the disc can reveal whether the forty-four elements of nature present in the human body are in excess or are deficient. $29.00.
SPECTRUM MIRROR: A many-amplified pendulum containing radium, a wave-guide, and a silver mirrored surface to reflect off unwanted vibes including negative thoughts. Excellent for the professional and the beginner. As a gift it is very beautiful. $18.00.
MAGNETRON: This instrument is a broadcast healer and direction finder, allowing you to send healing vibrations to anyone whose photograph you place at the center. Also a great gift for the organic gardener to protect plants from insects. $28.00.
SQUARE BALANCE EQUALIZER: A tool for absorbing and balancing life-force. $28.00.
THE ELECTRIFIED PENDULUM: A very special pendulum in that it is constantly electrified by a p/n junction and so can be used in tuning your car, finding what the trouble is when your car suddenly breaks down. It is good for any electrical tuning, and charging or discharging yourself sexually, depending on what type of energy you need. $35.00.
TIME RECESSIONAL DISC: Enables a person with a Merkhet or Osiris pendulum to date and orientate ancient objects. $7.50.
TURENNE RULE: Reveals the 44 elements of nature which are present in the human body, excess or deficit of any of which will cause ill health. It is a remarkable instrument for finding out your true spiritual condition, and your physical and emotional state of being. A real bargain at $20.00.
RADIUM BLOCK, TRITIUM BLOCK: Are amplifiers for all divining instruments. Both are $23.00. The Radium Block is specially made to fit on the Turenne Rule and amplifies the field of the sample itself. The Tritium Block amplifies the diviner's reaction.
HEFIGAR: Amplifies the field so strongly that, using it, most people can become diviners. It is a virtually indispensible instrument to be used in conjunction with a pendulum or rod. $40.00. Name is abbreviation for "Hills End-Fire Intensifying Guide and Radiator."

These prices subject to change without notice.

UNIVERSITY OF THE TREES PRESS, P.O. Box 644, Boulder Creek, California 95006

Please send me the items I have checked above. I have figured the mailing charges as stated below and enclose $_____. Send check or money order, no cash or C.O.D.'s please. California residents add 6% sales tax. Foreign Mail Rates: Please add $2.50 to listed rates.

MAILING & HANDLING CHARGES
Add the charges indicated to your order to cover mailing. No stamps or C.O.D.'s please.
If your order is:

up to $6. add	$ .50
$6.01-$20. add	$1.00
$20.01-$40. add	$1.50
$40.01-$60. add	$2.00
$60.01-$80. add	$2.50
$80.01-$100. add	$3.00

et cetera.

Name..
Street Address..
P.O. Box..
City................ State........... Zip........
Phone...